Firefighter Fatalities
in the
United States
in 1995

Prepared for

United States Fire Administration
Federal Emergency Management Agency
Contract No. EMW-95-C-4713

Prepared by

TriData Corporation
1000 Wilson Boulevard
Arlington, Virginia 22209

August 1996

ACKNOWLEDGMENTS

This study of firefighter fatalities would not have been possible without the cooperation and assistance of many members of the fire service across the United States. Members of individual fire departments, chief fire officers, the National Interagency Fire Center, US Forest Service personnel, the US military, the Department of Justice, and many others contributed important information for this report.

This analysis was conducted by TriData Corporation of Arlington, Virginia, for the United States Fire Administration under contract EMW-95-C-4713.

The ultimate objective of this effort is to reduce the number of firefighter deaths, through an increasing awareness and understanding of their causes and how they can be prevented. We acknowledge that firefighting, rescue and other types of emergency operations are essential activities in an inherently dangerous profession, and that tragedies will occur from time to time. This is the risk all firefighters accept every time they respond to an emergency incident; however, the risk can be greatly reduced through efforts to increase firefighter health and safety. This report is dedicated to those firefighters who have made the ultimate sacrifice in 1995. May the lessons learned from their passing not go unheeded.

BACKGROUND

For two decades, the United States Fire Administration (USFA) has kept track of firefighter fatalities and conducted an analysis of the fatalities that occur each year. Through the collection of information on the causes of firefighter deaths, the USFA is able to focus on specific problems and direct efforts towards finding solutions to reduce the number of firefighter fatalities in the future. This information is also used to measure the effectiveness of current efforts directed toward firefighter health and safety.

The USFA also maintains a list of firefighter fatalities. Their next of kin are invited to the annual Fallen Firefighter Memorial Service, which is held at the National Fire Academy in Emmitsburg, Maryland every fall.

INTRODUCTION

The specific objective of this study was to identify all of the on-duty firefighter fatalities that occurred in the United States in 1995, and to analyze the circumstances surrounding each occurrence. The study is intended to help identify approaches that could reduce the number of deaths in future years. In addition to the 1995 findings, this study includes a special analysis of the use of personal alert safety devices at fatal structure fires and a special report on several fatalities that occurred during technical rescue operations.

This report continues a series of annual studies by the US Fire Administration of firefighter fatalities in the United States.

Who Is a Firefighter?

For the purpose of this study, the term *firefighter* covers all members of organized fire departments, including career and volunteer firefighters; full-time public safety officers acting as firefighters; state and federal government fire service personnel; including wildland firefighters; and privately employed firefighters, including employees of contract fire departments and trained members of industrial fire brigades, whether full or part-time. This also includes contract personnel working as firefighters or assigned to work in direct support of fire service organizations.

Under this definition, the study includes not only local and municipal firefighters, but also seasonal and full-time employees of the United States Forest Service, the Bureau of Land Management, the Bureau of Indian Affairs, the Bureau of Fish and Wildlife, the National Park Service, and state wildland agencies. It also includes prison inmates serving on firefighting crews; firefighters employed by other governmental agencies such as the United States Department of Energy; military personnel performing assigned fire suppression activities; and civilian firefighters working at military installations.

What Constitutes an On-Duty Fatality?

On-duty fatalities include any injury or illness sustained while on-duty that proves fatal. The term *on-duty* refers to being involved in operations at the scene of an emergency, whether it is a fire or non-fire incident; being en route to or returning from an incident; performing other officially assigned duties such as training, maintenance, public education, inspection, investigations, court testimony and fund-raising; and being on-call, under orders, or on stand-by duty, except at the individual's home or place of business.[1]

These fatalities may occur on the fireground, in training, while responding to or returning from alarms, or while performing other duties that support fire service operations.

A fatality may be caused directly by accident or injury, or it may be attributed to an occupational-related fatal illness. A common example of a fatal illness incurred on duty is a heart attack. Fatalities attributed to occupational illnesses would also include a communicable disease contracted while on duty that proved fatal, where the disease could be attributed to an occupational exposure.

Accidents that claim the lives of on-duty firefighters are included in the analysis, whether or not they are directly related to emergency incidents. In 1995, this category includes a firefighter who died in a car accident while returning from a training class and a fire commissioner who died of a heart attack at a state fire association convention.

Injuries and illnesses are also included in cases where death is considerably delayed after the original incident. When the incident and the death occur in different years, the analysis counts the fatality as having occurred in the year that the incident occurred. For example, a firefighter died in 1995 of medical complications that resulted from an exposure to a hazardous material in 1987. Because his death was the result of the 1987 incident, this case was counted as a 1987 fatality for statistical purposes, and is not included in the 96 fatalities for 1995 that were analyzed in this

---

[1] A volunteer responding from home or work would be considered "on-duty" from the moment he or she is called to respond to an alarm, and would remain "on-duty" until they had returned from the alarm.

report. Since the death occurred in 1995, he will be included in the 1995 annual Fallen Firefighter Memorial Service at the National Fire Academy, and his name will be included on the list of firefighters who died in 1995.

There is no established mechanism for identifying fatalities that result from illnesses that develop over long periods of time, such as cancer, which may be related to occupational exposure to hazardous materials or products of combustion. It has proven to be very difficult over several years to fully evaluate occupational illness as a causal factor in firefighter deaths, because of the limitations in the ability to track the exposure of firefighters to toxic hazards and the potential long-term effects of such exposures.

Sources of Initial Notification

As an integral part of the ongoing program to collect and analyze fire data, the US Fire Administration solicits information on firefighter fatalities directly from the US fire service and from a wide range of other sources. These include the United States Fire Administration, and the Public Safety Officer's Benefit Program (PSOB) administered by the Department of Justice, the Occupational Safety and Health Administration (OSHA), the US military, the National Interagency Fire Center, and other federal agencies.

The Fire Administration also receives notification directly from fire departments, as well as from fire service organizations such as the International Association of Fire Chiefs (IAFC), the International Association of Fire Fighters (IAFF), the National Fire Protection Association (NFPA), the National Volunteer Fire Council (NVFC), state fire marshals, state training organizations, other state and local organizations, and fire service publications. The US Fire Administration also keeps track of fatal fire incidents as part of the Major Fire Investigations Project and maintains an ongoing analysis of data from the National Fire Incident Reporting System (NFIRS) for the production of *Fire in the United States.*

Procedure for Including a Fatality in the Study

After notification of a fatal incident is received from any source, initial telephone contact is made by the contractor with local authorities to verify the incident, its location and jurisdiction, and the fire department or agency involved.

Further information may be obtained from the chief of the fire department or his designee over the phone or by data collection forms, for both the deceased firefighter and the incident. After this information is obtained by the contractor, a determination is made as to whether the individual qualifies as an on-duty firefighter fatality, according to the previously described criteria. Additional information may be requested, either by follow-up with the fire department directly, or through state vital records offices or other agencies. The final determination as to whether a fatality qualifies as an on-duty death for inclusion in the statistical analysis, and in the Fallen Firefighter Memorial Service, is made by the United States Fire Administration.

Information that is routinely requested includes NFIRS-1 (incident) and NFIRS-3 (fire service casualty) reports, the fire department's own incident reports and internal investigation reports, copies of death certificates or autopsy results, special investigative reports such as those produced by the USFA or NFPA, police reports, photographs and diagrams, and newspaper or media accounts of the incident. The same criteria have been used for this study as in previous annual studies that were conducted by the NFPA.

1995 FINDINGS

Ninety-six (96) firefighters died while on duty in 1995.[2] This is a decrease from last year's total of 104. The total of 96 fatalities is the third lowest number of fatalities recorded in the 19 years that this data has been collected and only the third time that the total has been less than 100 fatalities. The lowest years were 1992 with 75 fatalities and 1993 with 77 fatalities.

This continues the long-term downward trend (down 44 percent) of reduced fatalities that began in 1979 after a peak of 171 in 1978. Deaths in 1995 decreased approximately 8 percent from 1994. This decrease is shown in Figure 1.

The fatalities included 63 volunteer firefighters and 33 career firefighters (Figure 2). Among the volunteer firefighter fatalities, 57 were from local or municipal volunteer fire departments, four were part-time or seasonal members of wildland fire agencies, one was a member of an industrial fire brigade, and one was a member of a department's fire-police unit. Of the career firefighters who died, 28 were members of local or municipal fire departments, two were members of a military base fire department, and three were full time firefighters employed by wildland fire agencies. Ninety of the fatalities were men; six were women.

---

[2] A total of 104 firefighters died in 1995 who would meet the criteria for an on-duty fatality. Eight of these deaths were attributed to events that took place in earlier years, and are not included in this analysis. Three firefighters died of complications directly attributable to exposures to hazardous chemicals at emergency incidents, including one from chronic obstructive pulmonary disease (COPD) after an exposure to formic acid at an incident in 1987. The other two died from exposures to hazardous chemicals that occurred in 1989 and 1990. One firefighter died from HIV which he contracted during surgery that was necessitated by a 1980 on-duty injury. A firefighter who was paralyzed in an apparatus accident en route to an emergency in 1980 died from complications relating to his injuries. Two firefighters died while in comas, one resulting from an incident in 1979 and the other from a heart attack at a fire in 1994. One firefighter died from carbon monoxide poisoning that occurred at a 1994 incident.

Figure 1
On-Duty Firefighter Deaths 1977-1995

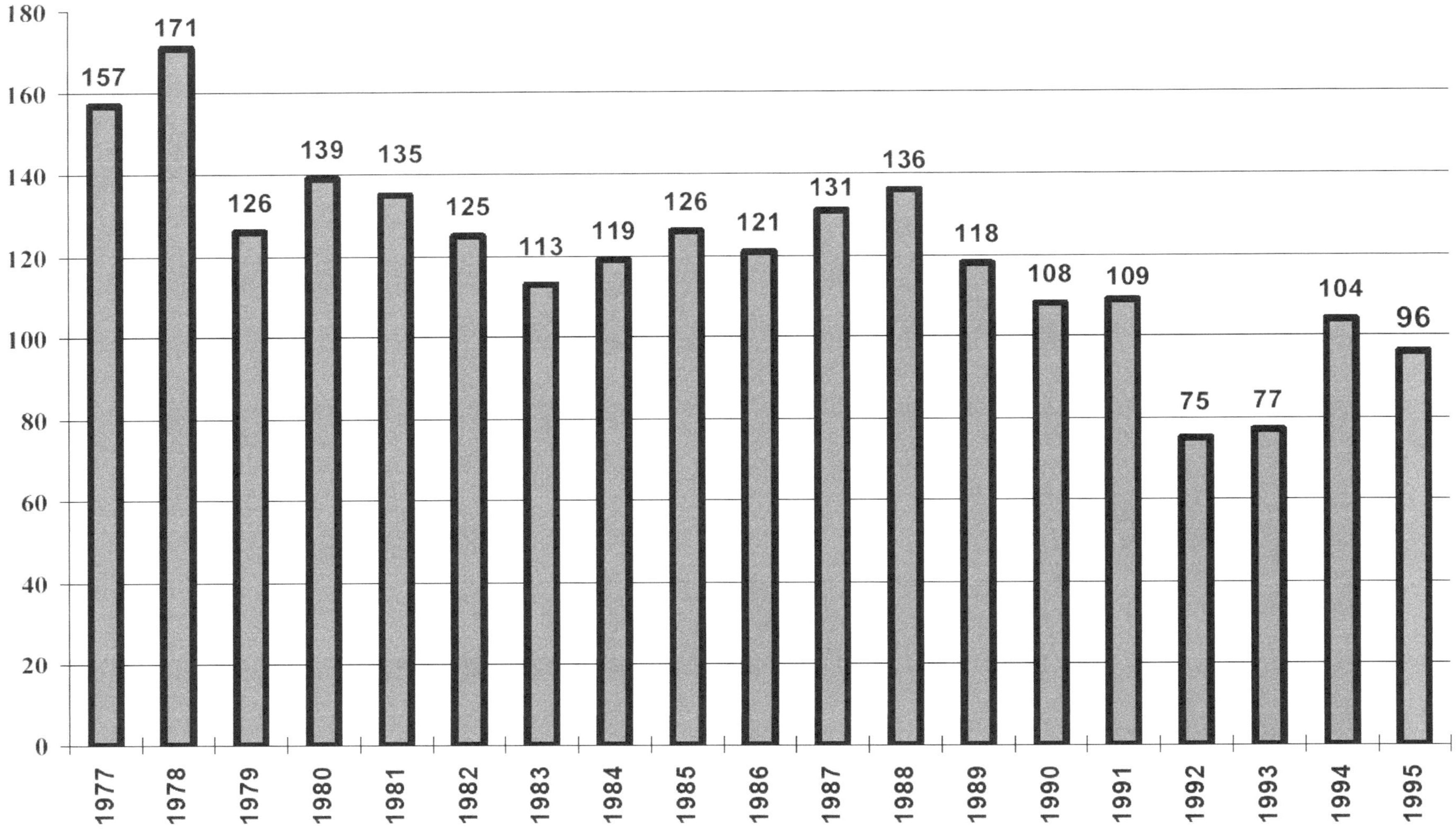

Figure 2
Career vs. Volunteer Deaths

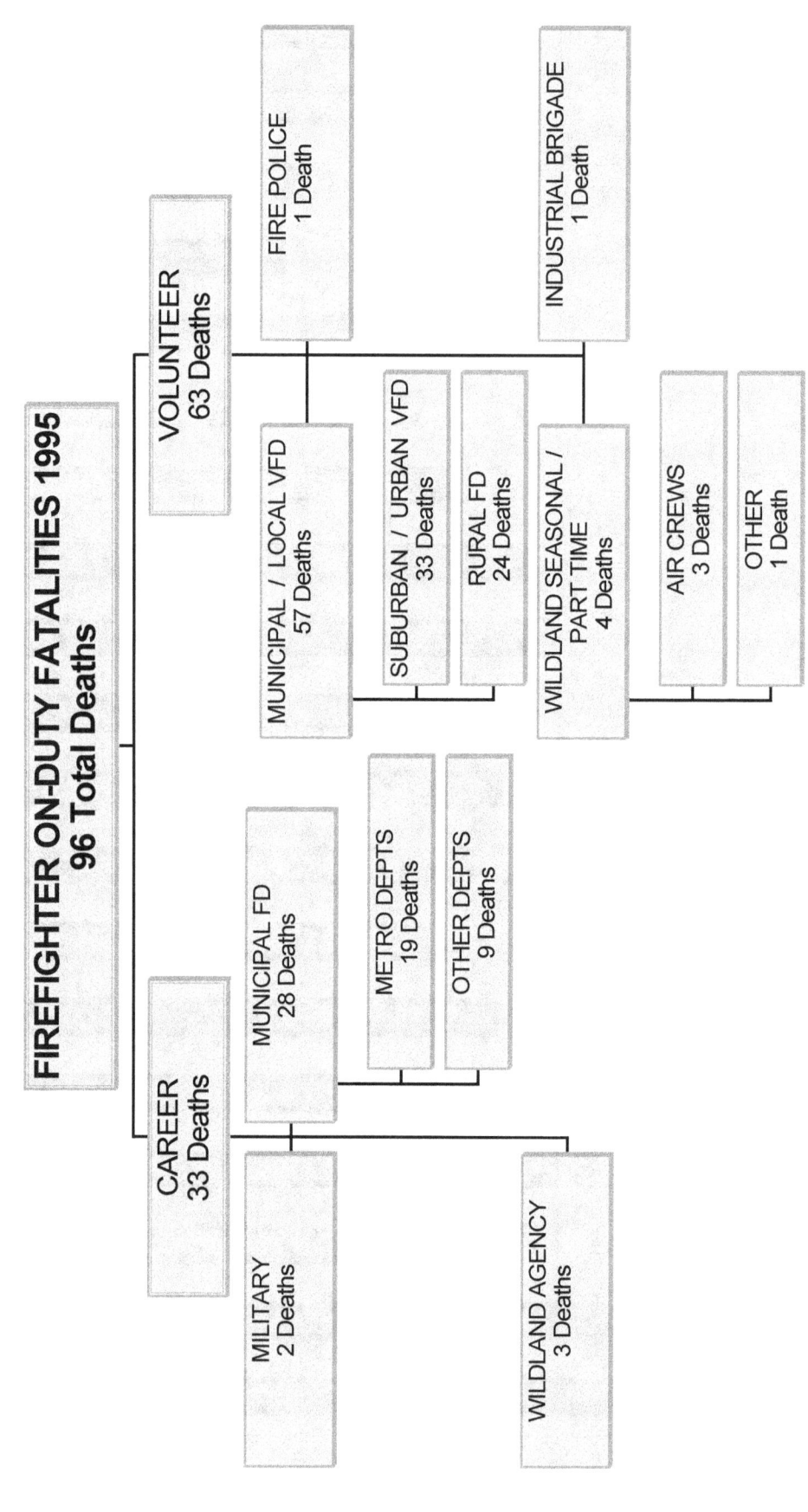

FIREFIGHTER ON-DUTY FATALITIES 1995
96 Total Deaths

CAREER
33 Deaths

VOLUNTEER
63 Deaths

MILITARY
2 Deaths

MUNICIPAL FD
28 Deaths

WILDLAND AGENCY
3 Deaths

METRO DEPTS
19 Deaths

OTHER DEPTS
9 Deaths

FIRE POLICE
1 Death

INDUSTRIAL BRIGADE
1 Death

MUNICIPAL / LOCAL VFD
57 Deaths

WILDLAND SEASONAL /
PART TIME
4 Deaths

SUBURBAN / URBAN VFD
33 Deaths

RURAL FD
24 Deaths

AIR CREWS
3 Deaths

OTHER
1 Death

The 96 deaths resulted from 85 incidents. There were seven multi-fatality incidents that resulted in 18 firefighter deaths. Four firefighters died in Seattle, Washington at a warehouse fire and three firefighters died in Pittsburgh, Pennsylvania at a house fire. Three pilots were killed in the mid-air collision of two planes returning from dropping retardant on a wildfire in California. Two military firefighters were killed in a fire at a civilian off-base oil refinery. Two firefighters were overrun by wildfire when their apparatus stalled. Two others died after their mini-pumper was hit by a train at a grass fire on a railroad right-of-way. Two firefighters died when a tanker overturned while en route to a fire.

The number of deaths associated with brush, grass or wildland firefighting dropped from the exceptionally high number of 38 in 1994 to 18 in 1995. Seven of these deaths resulted from three incidents: the aircraft crash, the stalled truck, and the truck that was hit by the train. Two firefighters were also killed in separate vehicle accidents en route to grass fires. One volunteer firefighter died of a stroke while responding to her station during the series of major wildfires that occurred on Long Island, New York.

Eight heart attack deaths were associated with brush, grass or wildland fire protection, including seven related to fire incidents and one that occurred while taking a stress test for a position with a state wildfire agency. Of the eighteen wildfire related fatalities, only four were full time career employees of wildland fire agencies.

Type of Duty

In 1995, 82 firefighter on-duty deaths were associated with emergency incidents, accounting for 85 percent of the 96 fatalities (Figure 3). This includes all firefighters who died while responding to an emergency, while at the emergency scene, or after the emergency incident. Non-emergency activities accounted for 14 fatalities. Non-emergency duties included training, administrative activities, or performing other functions that were not related to an emergency incident. Two firefighters who died of heart attacks while marching in parades are included in the non-emergency category.

A distribution of deaths by type of duty being performed is shown in Figure 4.[3] As in previous years, the largest number of deaths occurred during fireground operations, which accounted for 42 percent of the fatalities, down from 57 percent in 1994. There were 40 fireground deaths, 18 resulting from heart attacks on the scene, 15 from asphyxiation, four from trauma, and three from burn injuries.

The second largest category was responding to or returning from emergency incidents, which accounted for 29 deaths in 1995. This has been the second leading cause of deaths since 1993. Twelve firefighters suffered fatal heart attacks while responding or returning from emergency incidents. Seven firefighters were killed in fire apparatus accidents while en route to emergency incidents. All six of these accidents involved apparatus rollovers, and in at least three of these incidents the firefighters were not wearing seat belts. One death involved a firefighter who was killed when he was thrown off from the back step of an engine during an overturn accident.

Five firefighters were killed in accidents involving their personal vehicles while en route to emergency calls -- three of these were collisions with vehicles driven by other emergency responders en route to the same incident. Three pilots were killed when two aircraft collided while returning from a wildfire. One fire chief died while backing a fire engine into the station after a call, when he fell and received a fatal head injury. One firefighter had a stroke while responding to the fire station.

Thirteen deaths were related to activities at the scene of non-fire emergency incidents. Six firefighters died of heart attacks at EMS or rescue incidents. Five firefighters died during technical rescue incidents, including three who drowned while operating in swift moving flood waters and a fourth who died during a diving accident in a quarry. (Two of these deaths occurred during body recovering operations.) One industrial fire brigade member died during a confined space operation when he was overcome by an oxygen deficient environment and asphyxiated while attempting to rescue a worker from an excavation. One firefighter died in a helicopter crash while involved in a search and rescue operation and one firefighter was electrocuted by a downed power line.

---

[3] A summary of all the firefighter fatality incidents is included as Appendix A.

Figure 3

Firefighter Deaths While Performing Emergency Duty - 1995

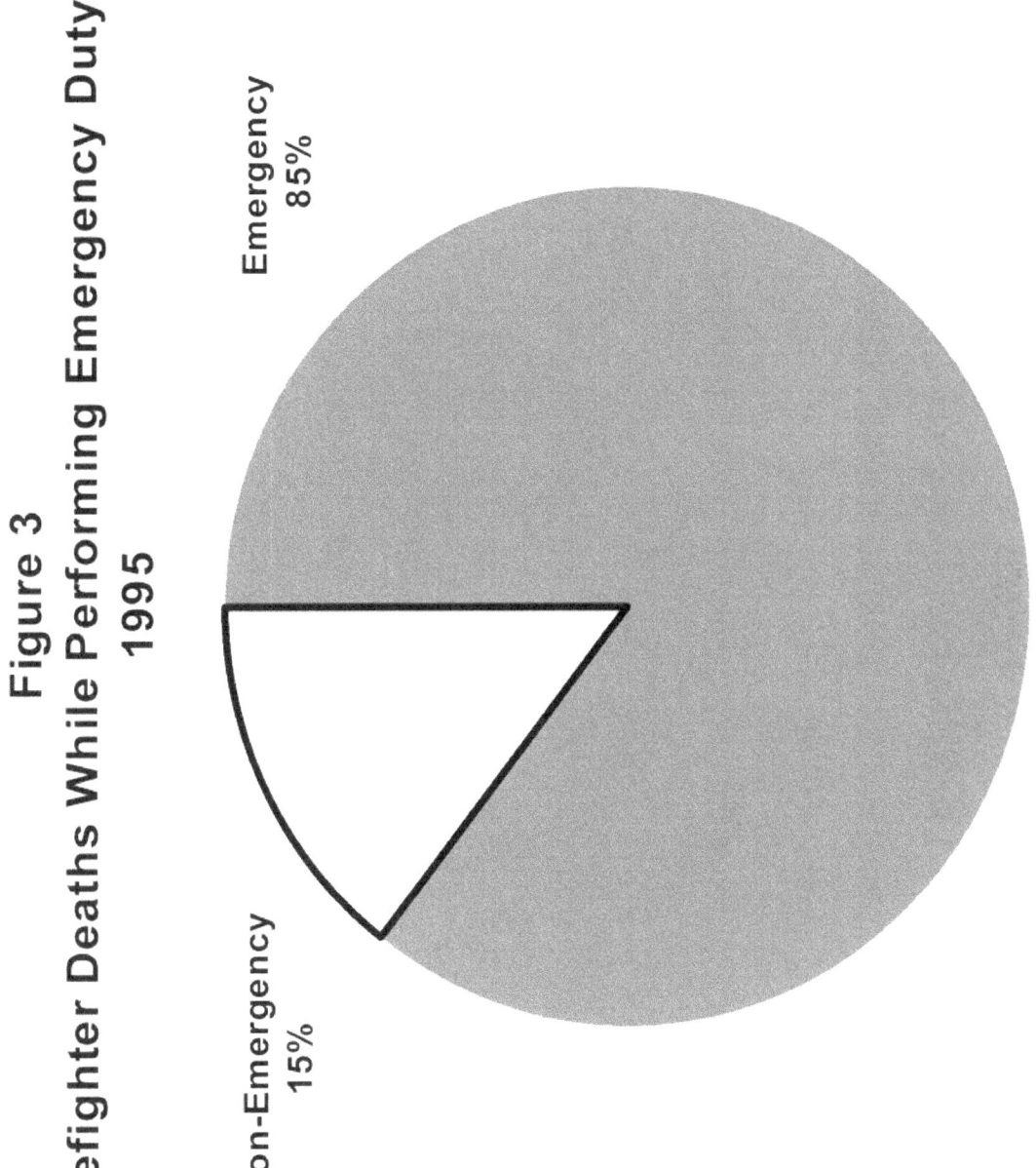

Emergency
85%

Non-Emergency
15%

Figure 4
Firefighter Deaths by Type of Duty - 1995

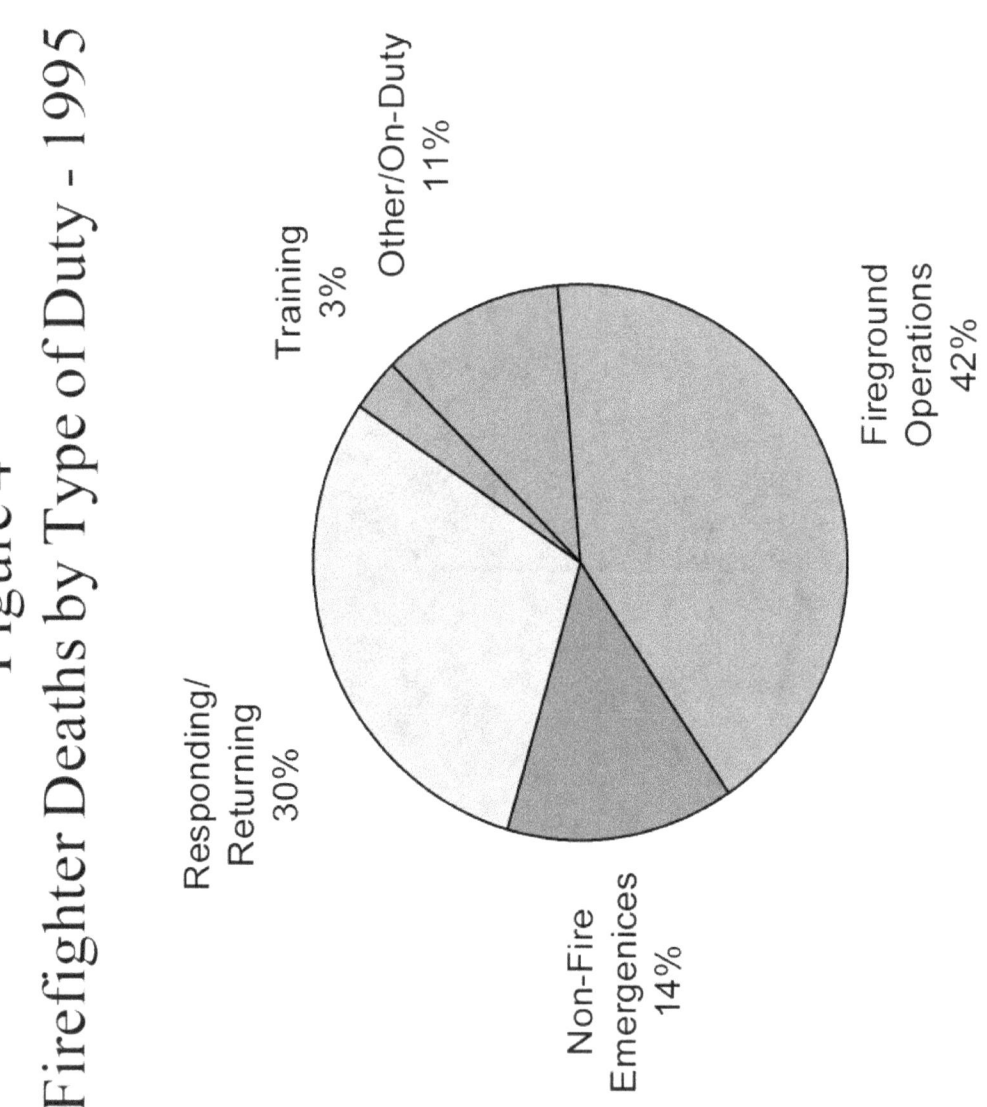

Eleven deaths occurred during non-emergency duty activities. These deaths include nine firefighters who died from heart attacks while on duty -- two at parades, two at fire department conferences, two while performing stress and agility tests, one while participating in a fire department fundraiser, one while conducting a fire inspection, and one while on duty in the fire station. One firefighter died from a seizure while on-duty as a dispatcher and one other firefighter was killed when his ambulance crashed during a routine drive.

Three deaths were attributed to training activities, including one death in an apparatus accident during driver training. One firefighter was killed in a car accident while returning from a paramedic class, and an instructor collapsed and died while teaching a confined space rescue course. There were no deaths associated with live fire training.

Cause and Nature of Fatal Injury or Illness

As used in this study, the term *cause* refers to the action, lack of action, or circumstances that directly resulted in the fatal injury, while the term *nature* refers to the medical nature of the fatal injury or illness, or what is often referred to as the physiological cause of death. Often, the fatal injury is the result of a chain of events, the first of which is recorded as the cause. For example, if a firefighter is struck by a collapsing wall, becomes trapped in the debris, runs out of air before being rescued, and dies of asphyxiation, the cause of the fatal injury is recorded as "struck by collapsing wall" and the nature of the fatal injury is "asphyxiation." Similarly, if a wildland firefighter is overrun by a fire and dies of burns, the cause of the death would be listed as "caught/trapped," and the nature if death would be "burns." This follows the convention used in NFIRS casualty reports, which are based on NFPA Fire Reporting standards.

Figure 5 shows the distribution of deaths by cause of fatal injury or illness. As in most previous years, the largest category is stress or overexertion, which was listed as the primary factor in 50 percent of the deaths, up from 35 percent last year. Firefighting has been shown to be one of the most physically demanding activities that the human body performs, and most deaths attributed to stress result from heart attacks. Of the 48 stress related fatalities in 1995, 46 firefighters died of heart attacks, one died of a stroke, and one died of a seizure. Eleven of the 48 deaths listed as stress-related occurred during non-emergency activities.

The second leading cause of firefighter fatalities was being struck by or coming in contact with an object. Of the 25 firefighters (26 percent) who died in these incidents, 15 were involved in vehicle accidents, four died in aircraft crashes, two were struck by vehicles while on emergency scenes, two were struck by a train at a fire scene, and one was struck by a collapsing wall. One firefighter was electrocuted when he came in contact with a downed electric power line.

The third leading cause of firefighter fatalities was being caught or trapped, which accounted for 20 deaths (21 percent). Five firefighters died as a result of becoming trapped by floor collapses, four in a warehouse fire and one in a tenement house. Four firefighters were trapped by rapidly changing fire conditions inside burning structures and one died at a residential fire when a garage door closed, trapping his crew inside. Two firefighters were overrun by a rapidly moving wildland fire when their truck stalled. Two others were caught in a boilover in a oil refinery fire. Three firefighters were caught by swift moving flood currents and drowned; two who were tied into ropes while attempting flood rescues and one who was trapped by a hydraulic (recirculating water current) at a body recovery.

Two asphyxiation deaths were attributed to exposure. One of these was a firefighter who died when he entered an oxygen deficient atmosphere without an SCBA (Self-contained Breathing Apparatus) or SABA (Supplied Air Breathing Apparatus). The second was caused by breathing a bad mixture of air from a SCUBA tank while engaged in a body recovery operation at 200 foot depth.

One firefighter died when he fell and struck his head after returning from a fire incident.

Figure 5
Firefighter Deaths by Cause of Fatal Injury

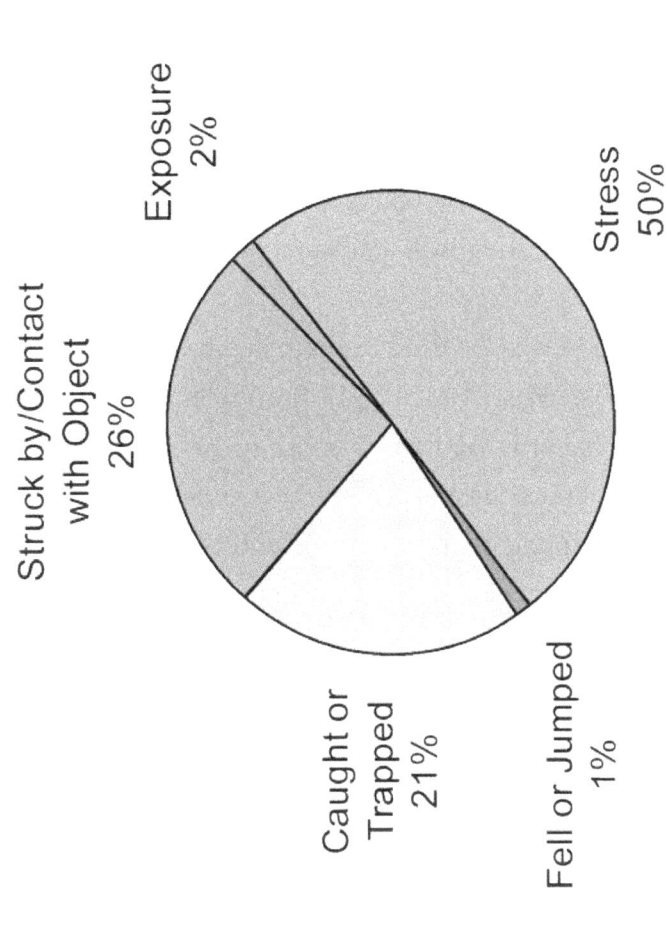

Struck by/Contact
with Object
26%

Exposure
2%

Stress
50%

Caught or
Trapped
21%

Fell or Jumped
1%

Figure 6 shows the distribution of the 96 deaths by the medical nature of the fatal injury or illness. Forty-six firefighters died of heart attacks in 1995, including at least 19 who were known to have high risk factors for heart attacks, including prior heart conditions, high blood pressure, obesity, or smoking.[4] Two of the heart attacks occurred during agility or stress testing. Eighteen firefighters suffered heart attacks at fire scenes and twelve suffered heart attacks en route to or returning from calls. Six heart attacks occurred at EMS or rescue incidents. Two heart attack deaths occurred at fire conferences, two while marching in parades, one while conducting a fire inspection, one while instructing a class, one while on duty in the station, and one at a fire department fundraiser.

Internal trauma was the second leading nature of death, responsible for 24 deaths. This total includes 17 firefighters who were involved in vehicle accidents and 4 in aircraft crashes. This category also includes 2 firefighters who were killed when they were struck by vehicles at emergency scenes, and one firefighter who died after being struck by a collapsing wall at a warehouse.

Asphyxiation was the third leading medical reason for firefighter deaths, responsible for 20 deaths. A total of 13 firefighter deaths resulted from carbon monoxide poisoning or inhalation of smoke or superheated gases during structural firefighting. All of these deaths occurred when the firefighters were caught and trapped by rapidly spreading fires or structural collapses. (Seven of these deaths occurred in only two incidents.) Two firefighters died of asphyxiation in a stalled brush truck when they were overrun by a wildfire. Three firefighters drowned in fast moving flood waters, one while attempting a body recovery in a creek, one while conducting a search for occupants of a vehicle caught in rising flood waters, and one after successfully rescuing an occupant of a vehicle trapped in flood water. One rescuer died of asphyxiation during a body recovery in a quarry, and one industrial firefighter was asphyxiated when he attempted a rescue in a oxygen deficient atmosphere without breathing apparatus.

---

[4] Autopsy results and medical records were not available for all of the heart attack victims.

Burn injuries claimed the lives of 3 firefighters. Two of these firefighters died of burns after being caught by a boilover at a oil refinery fire, and one died of burns after being caught in the basement of a tenement building when the floor collapsed beneath him.

One firefighter was electrocuted by a downed power line. One died of a stroke while en route to the station, and one died from seizure while on-duty.

Ages of Firefighters

Figure 7 shows the distribution of firefighter deaths by age and cause of death. Younger firefighters were more likely to have died as a result of traumatic injuries from an apparatus accident or after becoming caught or trapped during firefighting operations. Stress was shown to play an increasing role in firefighter deaths as ages increased. This is also reflected in Figure 8 which shows the distribution of deaths by age and nature. Trauma and asphyxiation were responsible for most of the deaths among younger firefighters, while heart attacks were much more prevalent among older firefighters. Twenty-six of the 31 firefighters who were over 50 years old and all 8 of the firefighters over 60 years old died from heart attacks.

Figure 6
Firefighter Deaths by Nature of Fatal Injury

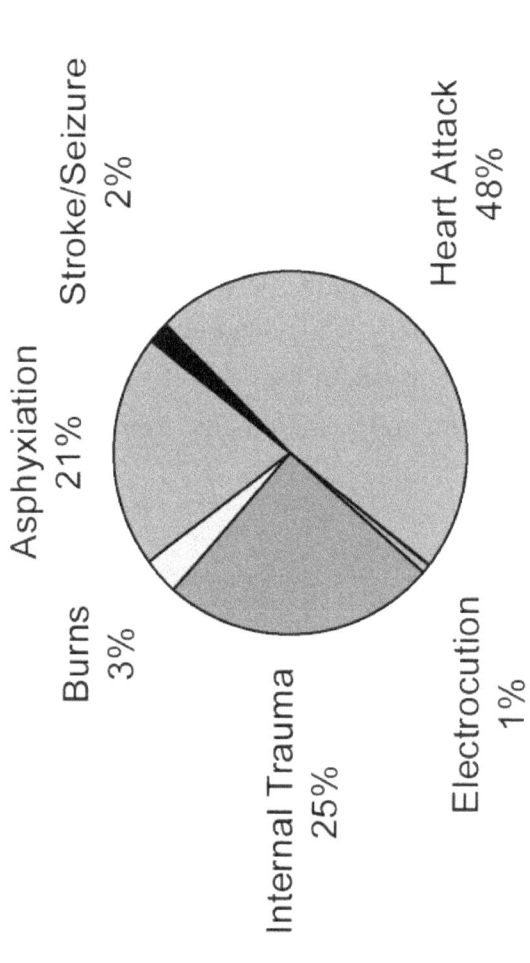

Figure 7

Firefighter Deaths by Age and Cause - 1995

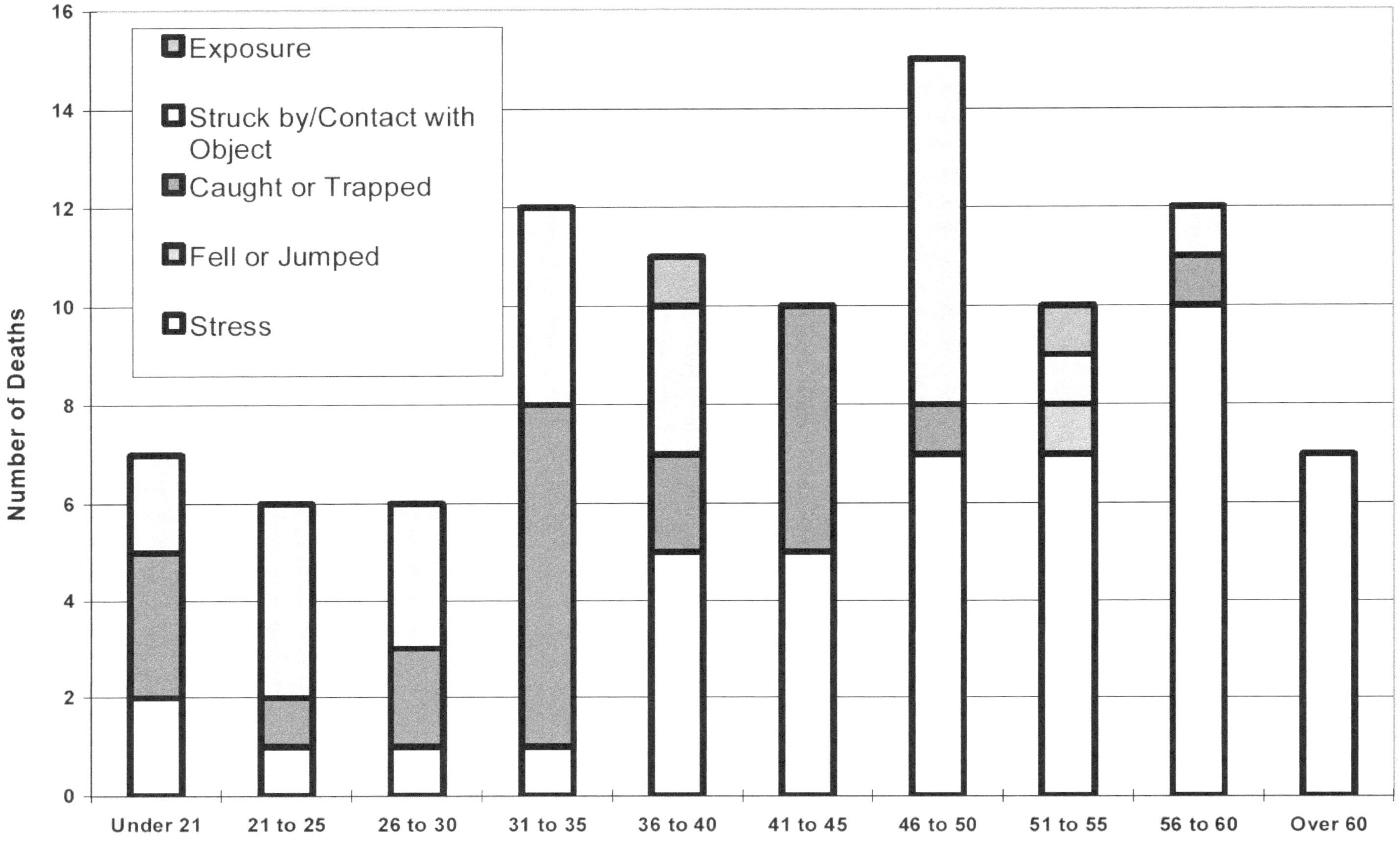

Legend:
- Exposure
- Struck by/Contact with Object
- Caught or Trapped
- Fell or Jumped
- Stress

Y-axis: Number of Deaths

X-axis categories: Under 21, 21 to 25, 26 to 30, 31 to 35, 36 to 40, 41 to 45, 46 to 50, 51 to 55, 56 to 60, Over 60

Fireground Deaths

Fireground deaths decreased by a third from 1994 to 1995, primarily due to the decrease in wildland firefighting deaths. [5] There were 40 fireground deaths in 1995. Figure 9 shows the distribution of fireground deaths by fixed property use.

Twenty-seven of the fireground deaths occurred at structure fires. As in most years, residential occupancies accounted for the highest number of these fireground fatalities with 18 deaths. Residential occupancies usually account for 70-80 percent of all structure fires and a similar percentage of the civilian fire deaths each year, 67 percent of the firefighter deaths in 1995 occurred in residential structures.[6] The frequency of firefighter deaths in relation to the number of fires is much higher for non-residential structures. Six firefighters died in storage occupancies which include warehouses and other storage facilities. One firefighter died of a heart attack at a recycling plant fire, one died of a heart attack at a fire in a commercial use building, and the type of structure was unreported at one other heart attack death. No firefighters died in public assembly occupancies in 1995.

Outdoor properties accounted for 13 deaths. Two firefighters died at an oil refinery fire. Ten firefighters died while engaged in grass, brush or wildland firefighting in 1995, down from 22 in 1994. One chief was killed when he was struck by a passing car during an auto fire.

---

[5] The fireground fatalities would be approximately the same as last year if the 14 deaths at Storm King Mountain are not included in the 1994 statistics.

[6] Complete NFIRS data for 1995 fire incidence was not available at the time of this report, but typically residential fires account for between 70 and 80 percent of all fatal fires each year.

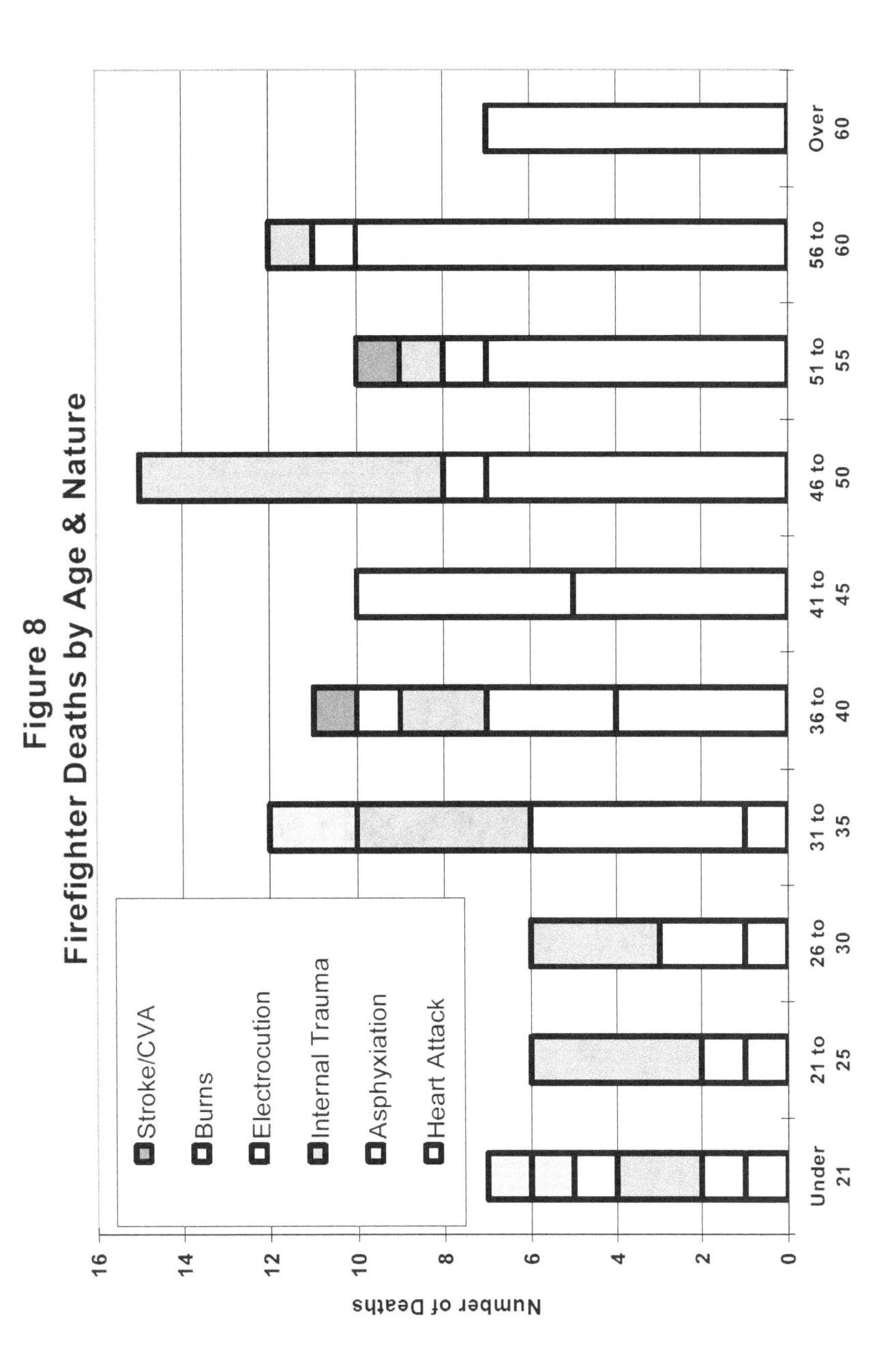

Figure 8
Firefighter Deaths by Age & Nature

Figure 9

Fireground Deaths by Fixed Property Use – 1995

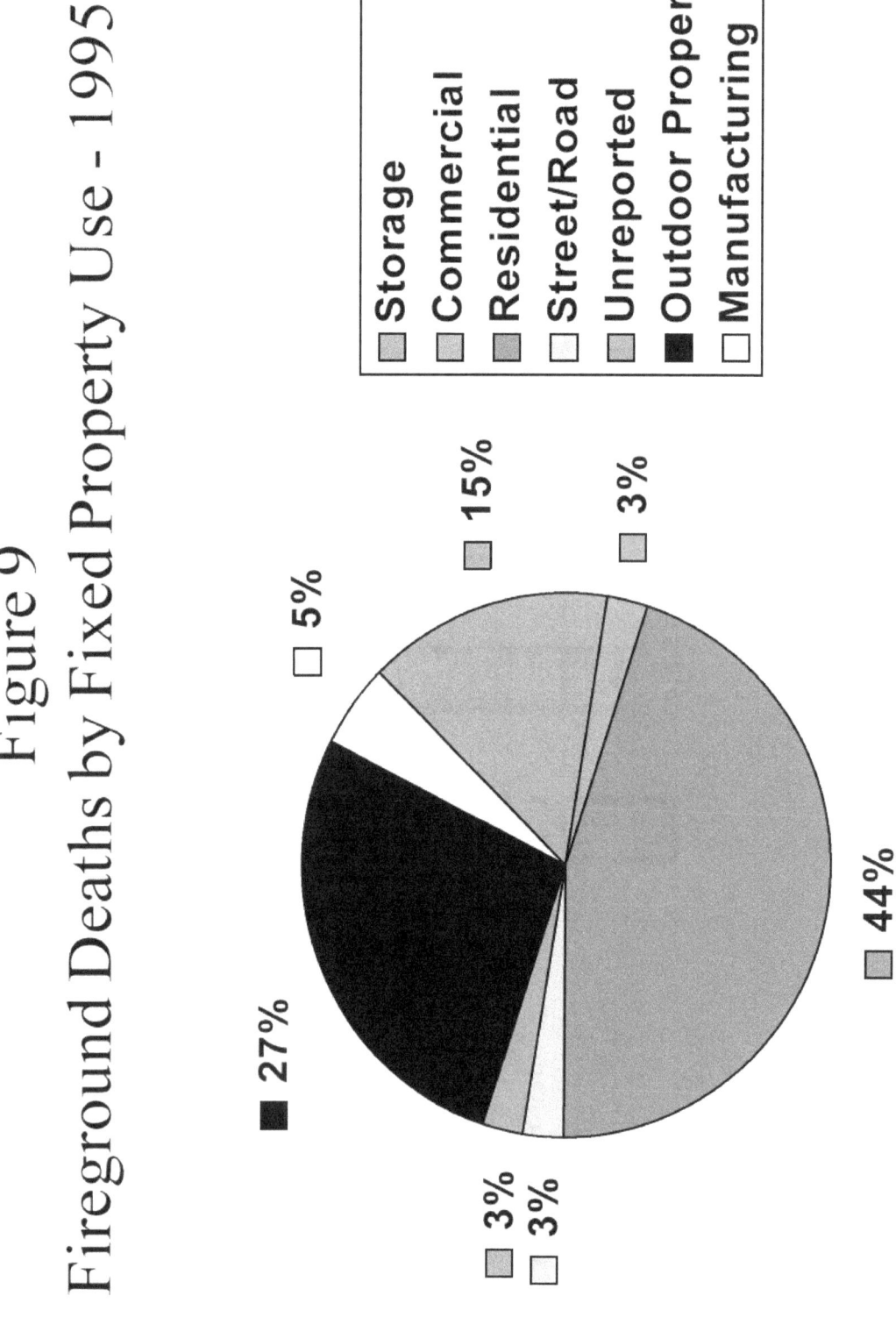

5%

15%

3%

44%

27%

3%
3%

Storage
Commercial
Residential
Street/Road
Unreported
Outdoor Property
Manufacturing

Figure 10 illustrates the activities the 40 deceased firefighters were engaged in at the time they sustained their fatal injuries or illnesses. There was a substantial increase this year compared to past years in the number of firefighters who died while engaged in traditional engine company duties of fire attack and advancing hose lines. Twenty-three firefighters died while performing these fireground operations, including 9 who died from asphyxiation after becoming trapped by rapid fire spread or structural collapse while advancing hose lines. Eight other firefighters suffered heart attacks while performing similar functions. Two firefighters died while attacking an oil refinery fire in a foam crash truck. Four other firefighters suffered heart attacks while performing water supply operations on the fireground.

Traditional truck and ladder company duties accounted for nine deaths. Search and rescue operations in burning structures were being conducted when five of these deaths occurred, a drop from 14 search and rescue deaths in 1994. Four of the 9 died of asphyxiation, all caught or trapped by rapidly spreading fires. One died of burns when the floor collapsed trapping him in a fully involved basement. Two firefighters died of heart attacks while performing ventilation at house fires. One firefighter died of heart attack during overhaul, and one died while forcing entry to a building when a brick wall collapsed, crushing him.

Three firefighters died of heart attacks while performing support functions or standing by on the fireground.

Cutting fire lines to contain grass, brush, and forest fires accounted for three firefighter fatalities. All three died as a result of heart attacks.

Two incident commanders suffered fatal heart attacks at fire incidents and one was struck by a vehicle.

Time of Alarm

The distribution of 1995 fireground deaths according to the time of day when the incidents were reported is shown in Figure 11. The highest number of fireground deaths occurred for alarms that were received between 0100 and 0300. The second highest number was between 1300-1500. There were no fireground deaths between the hours of 2100-2300.

Figure 10
Fireground Deaths by Type of Activity - 1995

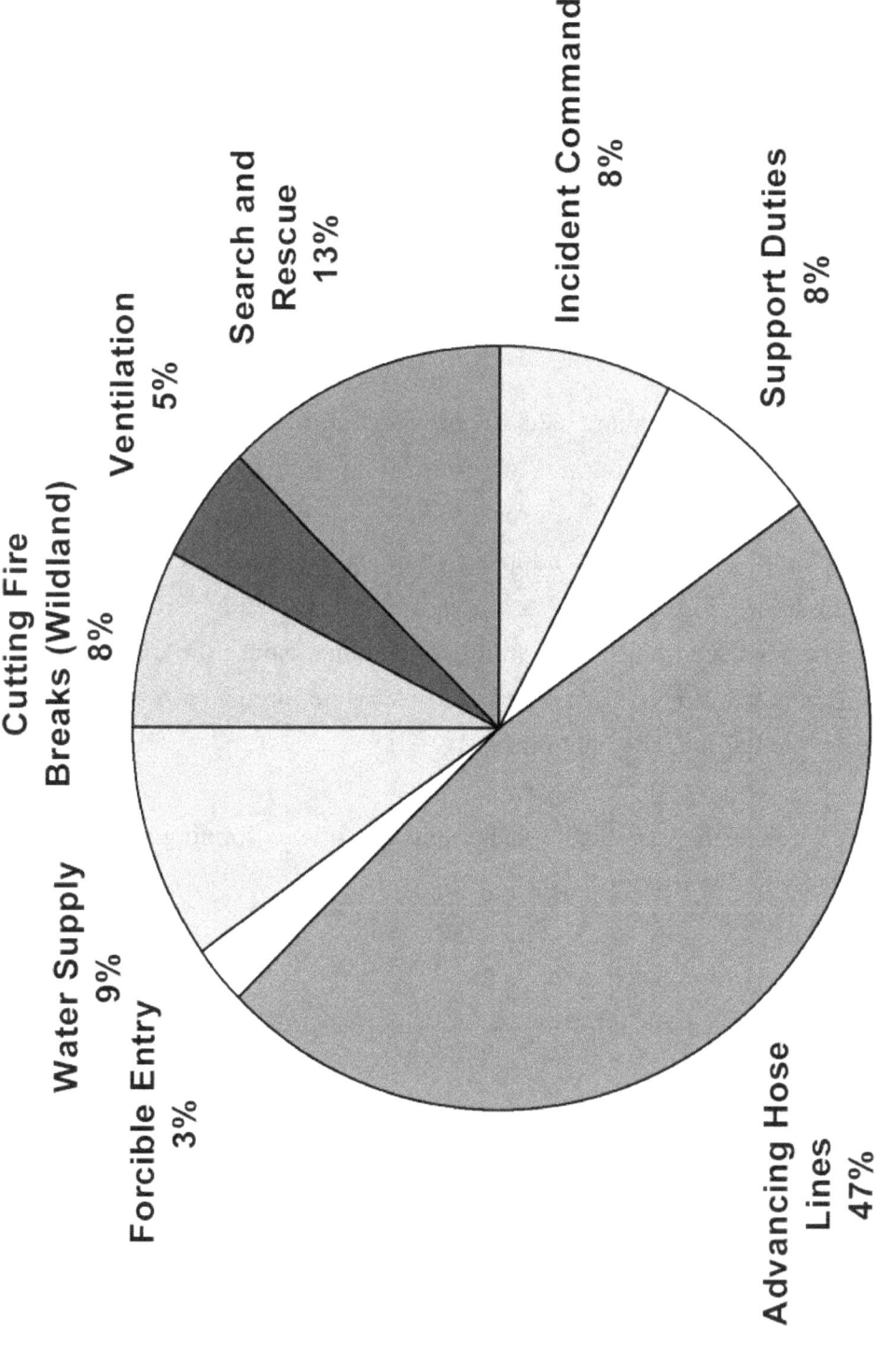

Search and
Rescue
13%

Ventilation
5%

Cutting Fire
Breaks (Wildland)
8%

Water Supply
9%

Forcible Entry
3%

Incident Command
8%

Support Duties
8%

Advancing Hose
Lines
47%

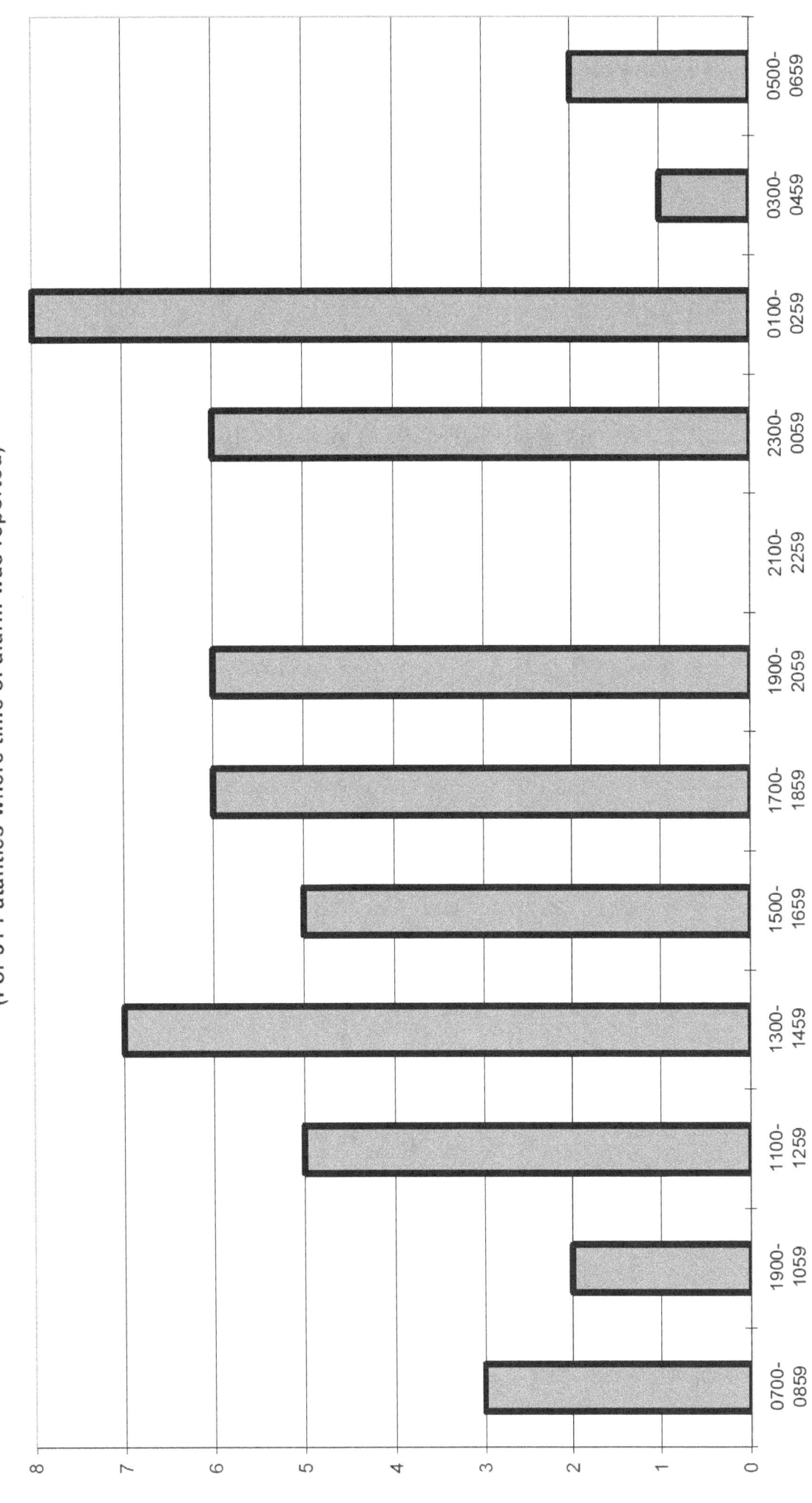

Figure 11
Time of Alarm - 1995
(For 51 Fatalities where time of alarm was reported)

Month of the Year

Figure 12 illustrates firefighter fatalities by month of the year. Firefighter fatalities peaked in June and July. Other high months were recorded in January and March. The late fall months into early winter (October, November, and December) were among the lowest months. (Conversely the number of residential fires peaked during the winter and was lowest during June and July.)

State and Region

The distribution of firefighter deaths by state is shown in Table 1.[7] Thirty-four states are represented on the list, led by New York with 13 deaths. Figure 13 shows the firefighter fatalities divided by region of the country and whether they were career structural, volunteer structural, or career or seasonal wildland firefighters.

---

[7] This list attributes the deaths according to the state where the fire department or unit is based, as opposed to the state where the death occurred. They are listed by those states for statistical purposes, and for the National Fallen Firefighters' Memorial at the National Fire Academy.

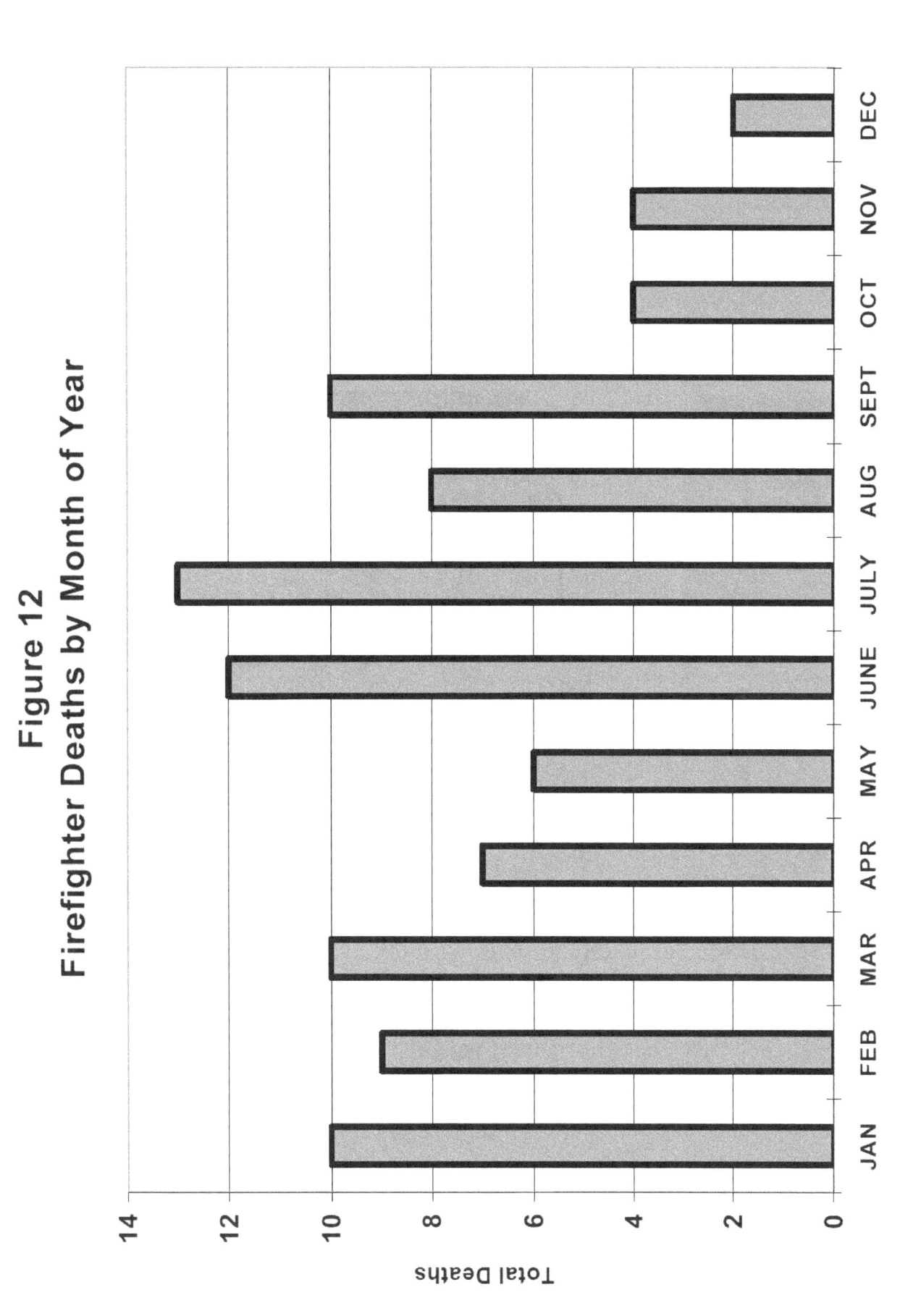

Figure 12

Firefighter Deaths by Month of Year

Figure 13
Firefighter Deaths by Region - 1995

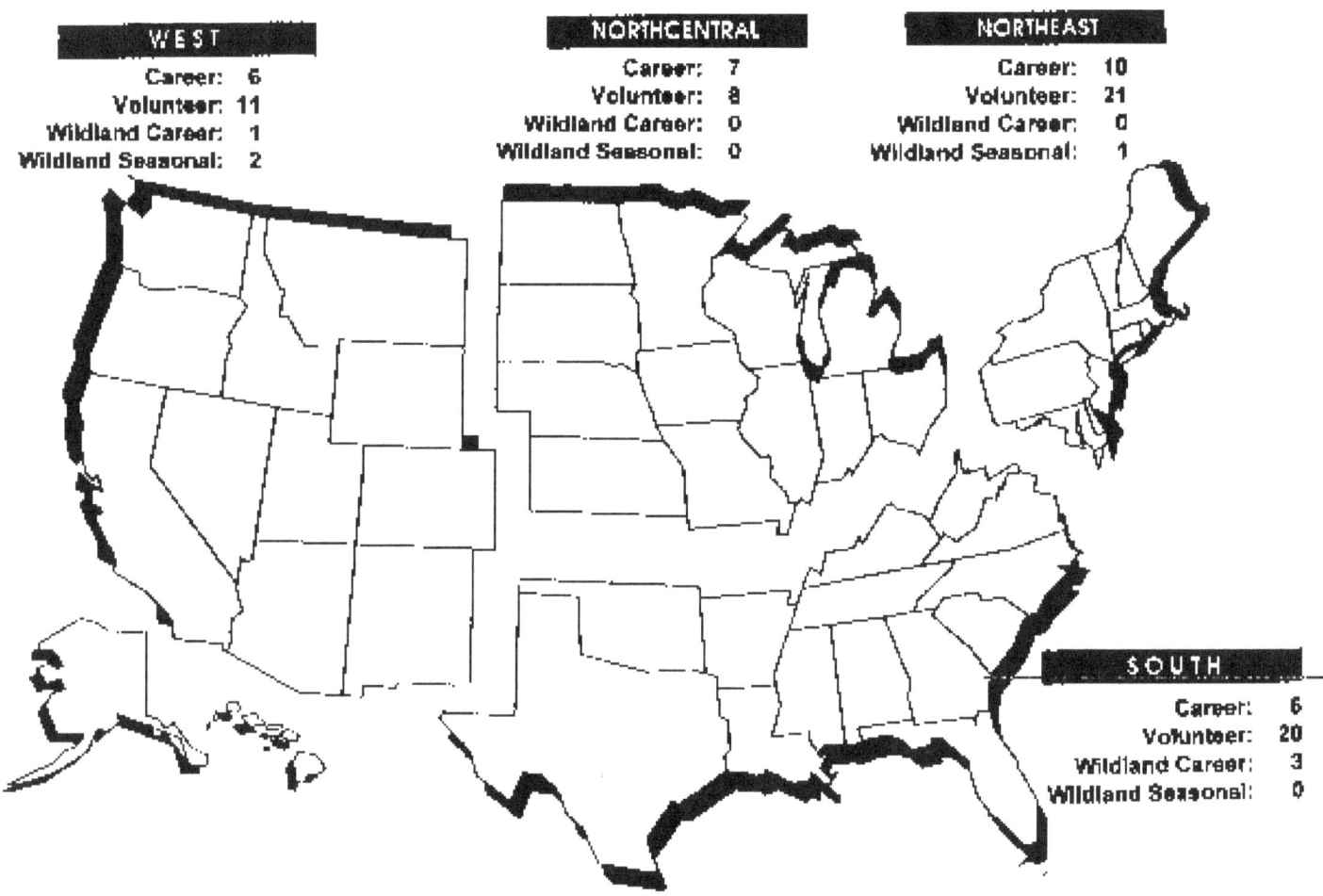

WEST
Career: 6
Volunteer: 11
Wildland Career: 1
Wildland Seasonal: 2

NORTHCENTRAL
Career: 7
Volunteer: 8
Wildland Career: 0
Wildland Seasonal: 0

NORTHEAST
Career: 10
Volunteer: 21
Wildland Career: 0
Wildland Seasonal: 1

SOUTH
Career: 6
Volunteer: 20
Wildland Career: 3
Wildland Seasonal: 0

Table 1.
1995 On-Duty Firefighter Fatalities

State	Number of Deaths	State	Number of Deaths
Alabama	2	New Mexico	1
Arkansas	2	New York	13
California	8	North Carolina	2
Colorado	1	Ohio	1
Connecticut	1	Oregon	1
Florida	1	Pennsylvania	8
Georgia	2	South Carolina	2
Hawaii	1	Tennessee	5
Idaho	2	Texas	6
Illinois	3	Virginia	2
Indiana	5	Washington	5
Kansas	2	West Virginia	2
Kentucky	1	Wisconsin	1
Maine	2		
Maryland	4		
Massachusetts	1		
Michigan	2		
Mississippi	2		
Missouri	1		
Nevada	1		
New Jersey	3		

Total: 96

Analysis of Urban/Rural/Suburban Patterns in Firefighter Fatalities

The US Bureau of the Census defines urban as a place having at least 2,500 population or lying within a designated urban area. Rural is defined as any community that is not urban. Suburban is not a census term but may be taken to refer to any place, urban or rural, that lies within a metropolitan area defined by the Census but not within one of the designated central cities of that metropolitan area.

Fire department areas of responsibility do not always conform to the boundaries used for the Census. For example, fire departments organized by counties or special fire protection districts may have both urban and rural sections. In such cases, it may not be possible to characterize the entire coverage area of the fire department as rural or urban, and firefighter deaths were listed as urban or rural based on the particular community or location in which the fatality occurred.

The following patterns were found for 1995 firefighter fatalities. These are estimates based upon data reported by the fire departments.

Table 2.

	Urban	Suburban	Rural	Federal or State Parks/Wildland	Total
Firefighter Deaths	33	33	24	6	96

Past analysis of urban, suburban and rural firefighter fatalities has shown that little correlation exists between the demographic nature of a department's area and firefighter fatalities. Experience has shown that training, use of proper equipment, and incident management have more of an impact on firefighter safety than geographic location.

FAILURE TO USE PASS DEVICES

Fourteen firefighters died in 1995 when they were caught or trapped in structural fires. Only two of these firefighters were found wearing personal alert safety system (PASS) devices that were in the "on" or "armed" positions. Eleven firefighters' PASS units were turned off when they died. In only one structural death was the status of the PASS unknown. In 1994, of six deaths where PASS information was available, only two firefighters died while wearing PASS devices that were turned "on" and four firefighters died with PASS units found turned off (the status of several others was not reported).

The eleven firefighters who died in 1995 without activating their PASS devices account for 79 percent of the structural firefighter deaths that were attributed to being caught or trapped by rapidly spreading fire conditions or structural collapses (Figure 14). In all but one of these cases asphyxiation was the primary cause of death, due to either toxic smoke inhalation or hypoxia after running out of air. All of these firefighters are believed to have been wearing breathing apparatus. It is possible that some of these firefighters might have been reached by fellow firefighters had their status and location been determined by PASS activation. PASS devices were not reported to be a factor in any of the heart attack deaths.

The use of PASS devices has gained widespread acceptance in the fire service since they were first developed ten years ago. The mandatory use of PASS devices was adopted as part of the NFPA 1500 *Standard on Fire Department Occupational Safety and Health Program* in 1987. Analysis of these recent structural fatalities clearly shows that many firefighters are failing to activate their PASS devices, even when they are trapped and able to initiate other emergency procedures.

Several reasons may exist for the failure of firefighters to arm their PASS devices. These include forgetting or intentionally not turning on the PASS when they activate their SCBA, turning off the PASS because many earlier models are prone to annoying false alerts, and failure to manually activate the PASS as part of their emergency survival procedures. Insufficient training in incorporating PASS activation into SCBA donning procedures and emergency procedures may be a contributing factor to some of these deaths. Traditional firefighter training is very specific and skills oriented, designed to

Figure 14
PASS Status of Firefighters Caught or Trapped During Interior Firefighting - 1995

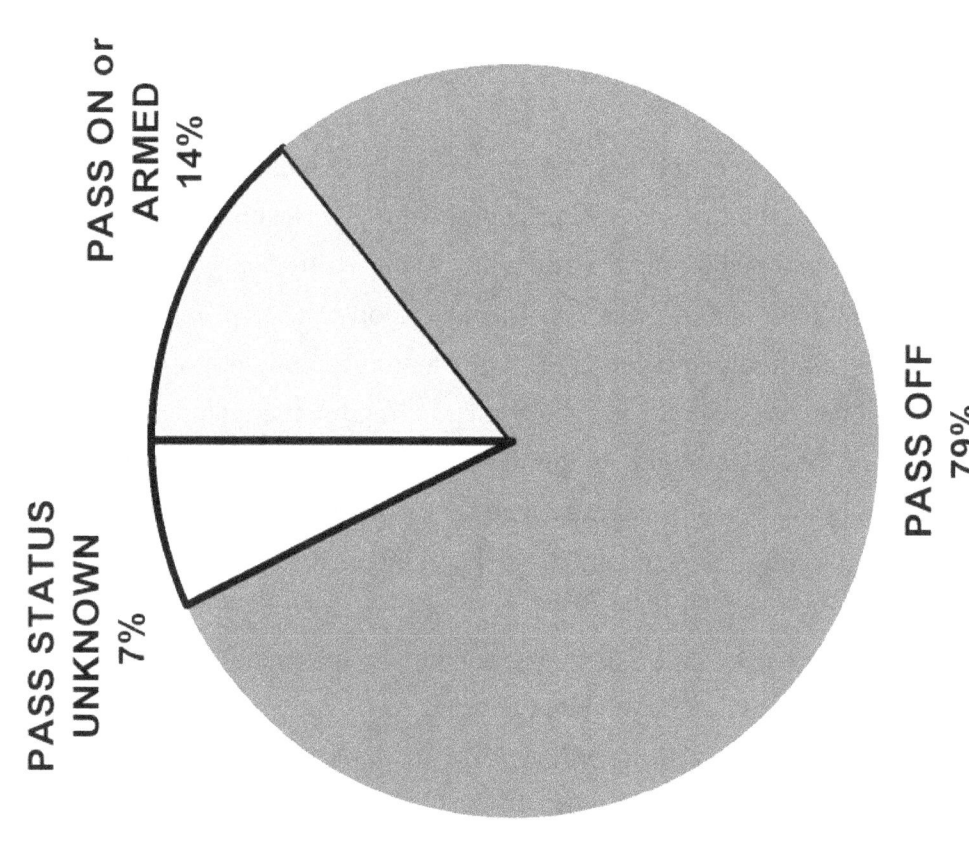

train new firefighters to act in an automatic and reactive fashion. Under stress, many firefighters, even more experienced ones, simply revert back to their original training which has become ingrained and instinctive, and may forget to arm PASS devices or activate them in emergencies. Some firefighters may not have received any specific re-training to make PASS use automatic and instinctive.

Several efforts may help increase the use of PASS devices. Initial firefighter training must incorporate the activation of PASS units every time an SCBA is turned on, to make the use of the devices instinctive. Manual PASS activation should also become a standard part of emergency survival procedures training for all trapped firefighters, so that this also becomes an automatic reflex. Thorough re-training should also take place for experienced firefighters, to overcome their old, learned behaviors.

Manufacturers have developed improved PASS devices in the last few years to meet revised NFPA standards and to reduce false activation. As these newer units replace older ones, firefighters should encounter fewer false alerts. Also, some manufacturers are developing integrated PASS devices that turn on automatically as part of SCBA activation, arming themselves as soon as the air bottles are turned on. This new technology should help make PASS use standard and universal as new SCBA systems become more widely used over the next decade.

The use of PASS devices must be emphasized in training and on the fireground. This should be an integral component of a fireground accountability system and a routine function for firefighters operating in hazardous areas.

DEATHS DURING TECHNICAL RESCUE OPERATIONS IN 1995

Deaths during technical rescue operations rose from none in 1994 to five in 1995. Four of these deaths were associated with water related emergencies, while one was associated with an attempted technical rescue in an excavation pit.[8] As indicated in this analysis report, only 14 of the 96 firefighter fatalities resulted from traumatic injuries during interior structural firefighting operations. Given that there are many more structural fires than technical rescue incidents every year, five deaths indicate a much higher level of risk during these incidents. (The exact death rate can not be determined and compared, because there is no standard reporting mechanism for all types of technical rescues.) The growing number of departments expanding into technical rescue operations makes these deaths more significant.

It is disturbing that all five technical rescue deaths involved rescuers who did not follow standard safety practices or adequately evaluate the level of risk they faced.

The field of technical rescue has grown rapidly, and industry consensus standards do exist for trench and confined space rescue, including mandatory OSHA regulations. Water rescue standards are currently under development by the NFPA, and coincide with industry consensus standards set by many state and private training agencies.[9]

The psychological profile of firefighters may contribute to the reasons why some of these deaths occur. Firefighters tend to be action and mission oriented persons. They are indoctrinated by fire department training and culture to overcome any obstacles to accomplish their tasks. This occasionally leads firefighters to improvise rescues in situations where they have not received any training. The widespread adoption of minimum training levels for firefighters has helped to reduce traumatic fireground deaths in the last two decades. Similar training is available for technical rescue, but has not been as widely adopted.

---

[8] One instructor died of a heart attack while teaching a confined space rescue course. His death is not included in this special analysis, which we have confined to emergency operations.

[9] An overview of applicable standards for technical rescue is presented in the *Technical Rescue Program Development Manual*, available on request from the United States Fire Administration.

Figure 15
Technical Rescue Deaths Versus Traumatic Deaths During Interior Structural Firefighting (19 Fatalities)

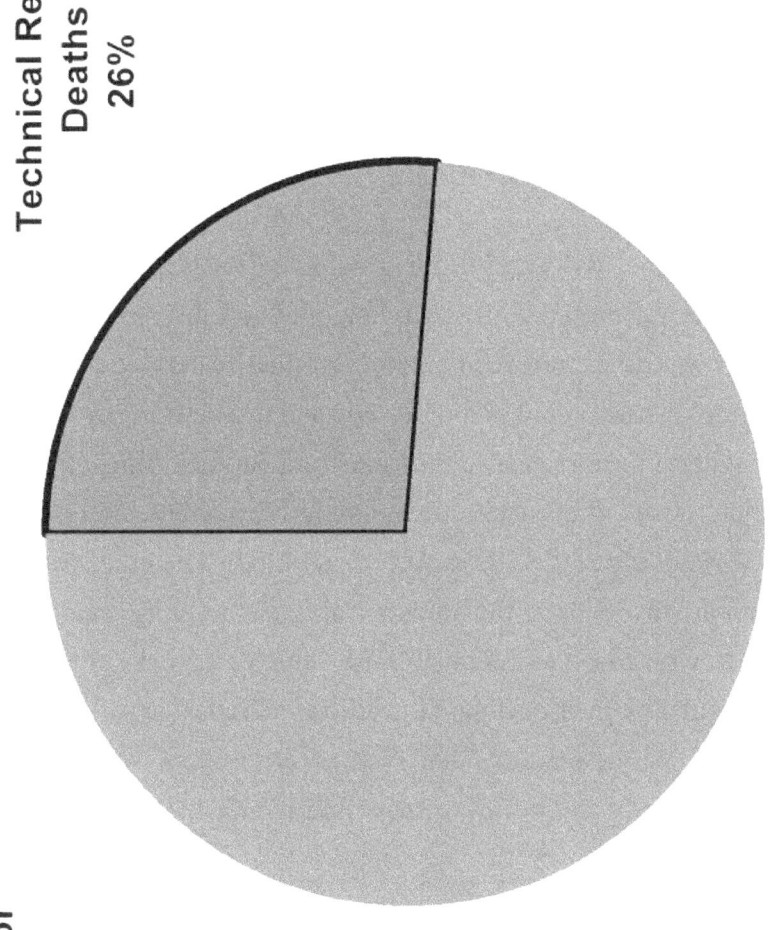

Technical Rescue
Deaths
26%

Traumatic Interior
Firefighting
Deaths
74%

All of the technical rescue deaths in 1995 involved action oriented rescuers in high risk situations attempting to overcome obstacles or unusual situations. Two of the five deaths involved body recovery operations, where the time pressure to take immediate actions is not present.

An analysis of the technical rescue deaths reveals common flaws in both risk assessment/evaluation and in tactical decisions that were made in the rescue efforts. Three of the deaths occurred during swiftwater and flood emergencies. Record flooding, especially in the Southeast United States, led to many flood related emergencies in areas where these types of incidents are unusual, and caused many fire departments to become involved in rescue incidents for which they were not trained.

In one swiftwater death, a volunteer assistant fire chief (who was also a career firefighter) drowned while searching automobiles along a flooded highway. The officer was an experienced member of the fire service who was a qualified state fire training instructor and had received rope rescue training, however, he was wearing full firefighting turnout gear in the water and had tied himself into a rope attached to a fire vehicle. Both of these practices should be avoided at all moving water emergencies. He was not wearing a life jacket of any kind. The chief waded into the water and was swept downstream by the quickly rising fast moving water and held underneath by the rope to which he was attached. The vehicles turned out to have been previously abandoned by their occupants, who were already in safe locations at the time of the incident. At the same incident, two other rescuers were thrown from a rescue boat and swept downstream, but were successfully rescued.

In a second swiftwater rescue fatality, a 34 year old volunteer firefighter tied himself to a rope with two other firefighters and waded out to a car which was being swept off a road into a swollen creek bed. As the firefighter reached the car, he was able to pull the driver from the front seat just as the car was swept into the fast moving water of the creek. As the force of the water increased, the firefighters and the driver were all swept downstream. A dive team member on the shore jumped into the water and was able to reach the driver. They were both swept downstream and later rescued by a boat. All three firefighters were held downstream by the static rope while shore personnel tried to pull them out of the water. The firefighter who had attempted the initial rescue of the victim became entangled in a guide wire to a telephone pole as he was pulled from the water, and drowned. Again, lack of training in the rescue environment, failure to properly size-up hazards, and choosing improper equipment and improper techniques for the situation led to a rescuer's death.

In both of these swiftwater drownings, the deaths occurred despite the use of standard fireground incident command procedures and a designated safety officer. However, the command structures lacked the necessary technical expertise to evaluate the special hazards posed in the swiftwater rescue environment.

A third swiftwater rescue death occurred when a forty year old firefighter drowned while attempting a body recovery in a creek. He and two other rescuers became trapped in a naturally occurring hydraulic below a falls. The two other firefighters were rescued and survived. All three personnel were members of a fire department dive team, and the firefighter that drowned was an instructor for dive, ice and water rescue, yet failed to assess the level of risk involved with the swiftwater body recovery operation. The rescuers were reportedly familiar with the dangers of hydraulics below man-made low-head dams, but did not realize that this phenomenon also occurs naturally as the result of water moving over any submerged object. The firefighter died while attempting to recover the body of a canoeist who had drowned in the same spot the day before. Recovery efforts had been suspended the day before due to dangerously high water levels.

The three personnel had formulated a plan to wade into the water below the falls wearing wet suits and buoyancy compensators. After a discussion, they decided to hold on to a safety rope with their hands instead of tying themselves into the line, for fear that if one of them got into trouble, the others would be unable to escape. This indicates that the divers were aware of a risk, and were attempting to improvise a plan. When the firefighters did wade into the water, they were drawn into the hydraulic and lost their grip on the safety line. One firefighter was able to rescue himself and a second was retrieved by a rescuer on shore with a pike pole, while the third firefighter was trapped under water and drowned before he floated free. A lack of understanding of the hazards, risks, and techniques for fast moving water resulted in a preventable tragedy.

All three of these deaths could have been prevented with proper training, proper equipment, and with an accurate assessment of the hazards to the rescuers.

A fourth water related death occurred when a rescue diver was asphyxiated in a quarry while diving to a depth of over 200 feet. It was determined that he had the wrong mixture of oxygen and nitrogen in his SCUBA tank. Deep dives require a special mixture of air to prevent the build up of nitrogen and other molecules in body tissues. This can cause hypoxia leading to reduced level of consciousness, disorientation, or unconsciousness in divers. The dive was part of a body recovery effort to find a drowning victim.

The fifth technical rescue death occurred when a member of an industrial fire brigade died of asphyxiation while attempting to rescue three unconscious workers who were in a pit that was being excavated as part of a construction project. The three workers had been overcome by an oxygen deficient atmosphere created when argon, a heavier than air gas, was accidentally pumped into the pit, displacing the oxygen. The fire brigade member entered the pit alone, without breathing apparatus, in an attempt to aid the workers and was quickly overcome from the lack of oxygen. A local technical rescue team responded to the plant and, using breathing apparatus removed the firefighter and the workers. The firefighter and two of the three workers died. Failure to size-up the hazard and risks involved in this rescue effort contributed to the firefighter's death.

Many technical rescues are inherently high risk incidents. These incidents are infrequent, many involve highly volatile environments, and often require specialized training and equipment and involve complicated techniques. All these technical rescue deaths might have been prevented with more thorough awareness training in the hazards of the rescue environment and a more complete size-up and analysis of the risks. The size-up must consider the limitations of the rescuers' training, equipment, and resources.

The increasing number of fire departments involved in providing technical rescue services must obtain the specific knowledge and skills to prevent repetitions of similar tragedies in the future. Confined space rescue requires training in accordance with OSHA regulations (CFR 1910.146). Underwater rescue and recovery requires thorough training in diving equipment and in the diving environment. Swiftwater rescue requires specialized training in hazard recognition and specific rescue techniques. Collapse and trench rescue require knowledge in shoring, building construction, and other disciplines.

An awareness of various technical rescue hazards should be incorporated into regular training for all first responder agencies and rescue services.

CONCLUSIONS

The analysis of firefighter deaths in 1995 indicates that the overall long term trend toward fewer firefighter fatalities is continuing. The 96 fatalities are the third lowest recorded, and only the third time the total number of fatalities has dropped below 100, all within the last four years.

Stress induced heart attacks continue to remain the number one reason for firefighter deaths.[10] Health and fitness programs should help to reduce these numbers in the long term. Better screening for high risk firefighters through medical exams may help to prevent some deaths in the short term, by identifying firefighters who are unfit for strenuous duty or may be at high risk for heart disease. Many factors that place firefighters at high risk are controllable, such as better nutrition or stopping smoking.

Two tragic fires resulted in the loss of seven firefighters. These and several other incidents reinforce the need for proper size-up, progress reports, as well as a working accountability system at all incidents to keep track of all personnel. An accountability system should track all members, including who and where they are, who they are working for, what they are doing, and how long they have been doing it. Accountability is an essential part of a fireground command system.

The institution of rapid intervention teams into emergency operations may help save more firefighters' lives. PASS devices must be used at all incidents, to improve the chances of being alerted to and locating a downed firefighter.

Responding to incidents continues to claim too many lives. After almost eliminating personal vehicle accidents as a cause of fatalities in 1994, several lives were lost in 1995. In at least three cases, responders crashed into each other while responding to the same incident. All fire departments should have a policy regarding driver training for responding to emergency incidents, and all drivers should use

---

[10] Figure 16 shows the relation between stress-related and action-related firefighter deaths.

caution when approaching intersections. Also, seatbelts can only save lives if they are worn; they were not used in several 1995 fatal accidents.

As in last years analysis, the 1995 statistics indicate that risk management, in the form of assessing firefighters health risks, sizing up fireground conditions, and evaluating hazards at special rescue scenes is the key to reducing on-duty firefighter fatalities even further.

Figure 16
Stress vs. Actions

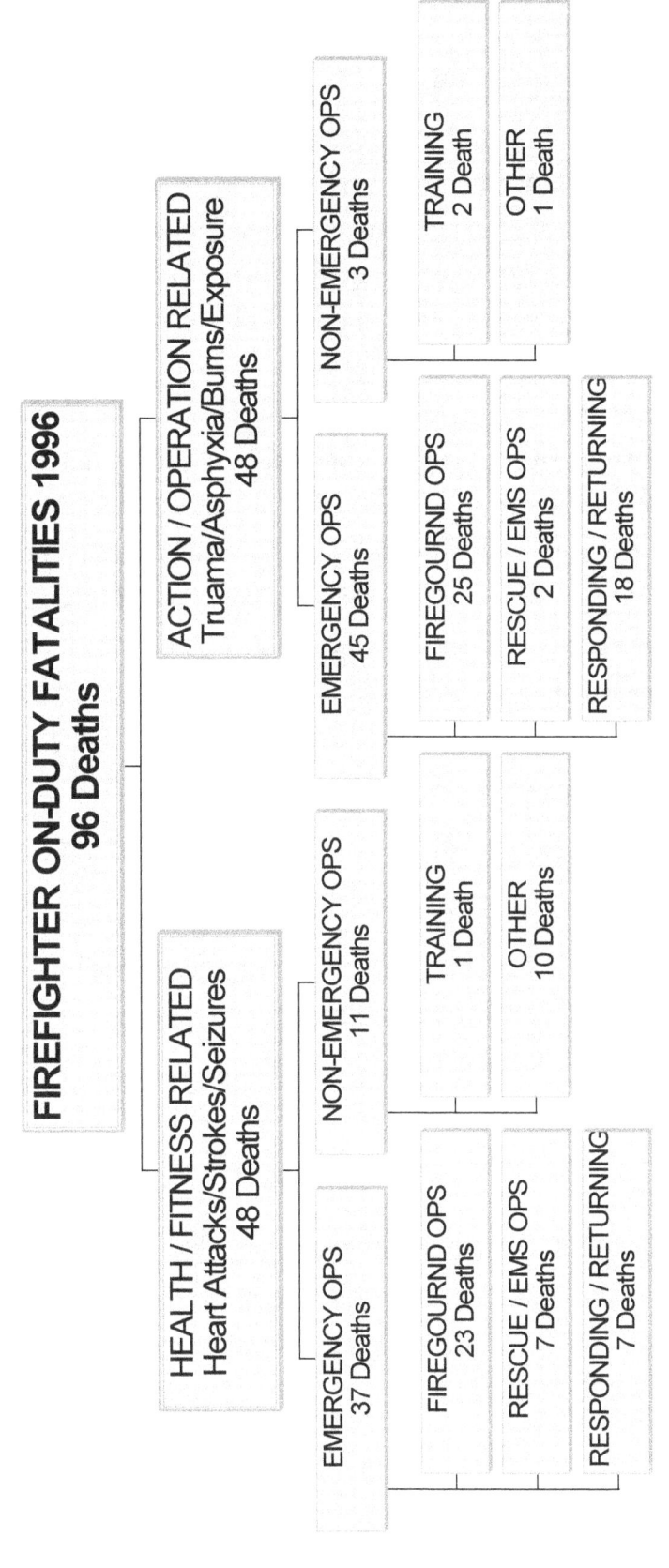

FIREFIGHTER ON-DUTY FATALITIES 1996
96 Deaths

HEALTH / FITNESS RELATED
Heart Attacks/Strokes/Seizures
48 Deaths

ACTION / OPERATION RELATED
Truama/Asphyxia/Burns/Exposure
48 Deaths

EMERGENCY OPS
37 Deaths

NON-EMERGENCY OPS
11 Deaths

EMERGENCY OPS
45 Deaths

NON-EMERGENCY OPS
3 Deaths

FIREGOURND OPS
23 Deaths

RESCUE / EMS OPS
7 Deaths

RESPONDING / RETURNING
7 Deaths

TRAINING
1 Death

OTHER
10 Deaths

FIREGOURND OPS
25 Deaths

RESCUE / EMS OPS
2 Deaths

RESPONDING / RETURNING
18 Deaths

TRAINING
2 Death

OTHER
1 Death

APPENDIX: SUMMARY OF FATAL INCIDENTS IN 1995

Incident 1

On January 5, four members of the Seattle (WA) Fire Department died when a floor collapsed without warning during a commercial building fire. Lt. Walter Kilgore, Lt. Gregory Shoemaker, Firefighter James Brown, and Firefighter Randall Terlicker died when a modified and unprotected wood floor support failed under heavy fire conditions. Contributing factors to this incident included an unusual and complicated building configuration, companies entering the structure on different levels from different sides of the building, a lack of pre-fire plans, conflicting interpretations of observed fire conditions from different locations on the fireground, personnel not recognizing the significance of their own observations and operations with respect to the overall incident, a lack of progress reports that would have permitted the incident commander to reevaluate his attack plan, and inadequate information passed on to responding companies about an arson threat against the building. The cause of the fire was determined to be arson, and a suspect has been charged. [11]

Incident 2

On January 7, Firefighter Wilbur Pinnell of the Winchester (TN) Fire Department suffered a fatal heart attack while returning from extinguishing a garbage fire that had extended to a commercial occupancy.

Incident 3

On January 24, Forestry Technician Henry Frizzell of the Tennessee Division of Forestry suffered a fatal heart attack while returning from a fire.

---

[11] A detailed analysis of this incident was conducted by the United States Fire Administration as part of the Major Fires Investigation Project. The report, "Four Firefighters Die in Seattle Warehouse Fire," is available on request from the United States Fire Administration.

Incident 4

On January 26, Chief Lathan Grant Smith, Jr., of the East Providence (AL) Volunteer Fire Department died after suffering a heart attack while fighting a brush fire. Grant was a founder of the East Providence VFD and a career firefighter with the Talladega Fire Department.

Incident 5

On January 28, Firefighter Victor Melendy of the Stoughton (MA) Fire Department died when he was caught in a flashover while searching for victims on the third floor of a rooming house.

Incident 6

On January 31, Firefighters Marcus King and Jared Lee Wright of the Claude (TX) Volunteer Fire Department were killed when their fire apparatus was struck by a train while they were fighting a brush fire on a railroad right-of-way. Both firefighters suffered severe traumatic injuries and died several days later. Wright was 18 years old at the time of his death; King was 15 years old.

Incident 7

On February 2, Firefighter Ernestine Garcia of the Willard (NM) Fire Department died when she was thrown from a fire engine in a rollover accident while responding to a brush fire. The vehicle apparently hit a soft shoulder on the side of the road, crossed the road and rolled over. The driver, a 17 year old junior firefighter, was hospitalized for multiple injuries. He was permitted to drive to fires but prohibited from operating on the fireground until he turns 18. The weather was clear and there was no oncoming traffic at the time of the crash.

Incident 8

On February 8, Chief Glenn Scott of the Era (TX) Volunteer Fire Department died from a heart attack while returning from an emergency call.

Incident 9

On February 13, Firefighter Lisa Batten of the Gilmer County (GA) Fire Department died from traumatic injuries she received in a car accident while returning from a paramedic class.

Incident 10

On February 14, three Pittsburgh (PA) Fire Department firefighters died after a stairway collapsed, trapping them in the basement. Captain Thomas Brooks, Firefighter Patricia Conroy, and Firefighter Marc Kolenda died from asphyxiation when they ran out of air while operating a hose line in the basement. Investigations by the City of Pittsburgh and others after the fire indicated that problems with incident command and accountability were key factors contributing to the firefighters deaths. Other factors included a possible lack of crew integrity and a failure of the crew to take emergency survival actions that may have helped them escape. All of the deceased firefighters were wearing PASS devices that were found in the "Off" position. The fire was ruled an arson and a suspect has been arrested.[12]

Incident 11

On February 14 Firefighter Wendell Ayers of the Pacific Grove (CA) Fire Department collapsed and died of an apparent heart attack while attempting to assist in the rescue of two people on a yacht that had run aground.

Incident 12

On February 22, Firefighter Shawn O'Brien of the Franklin (ME) Volunteer Fire Department collapsed and died of a heart attack after a structure fire.

_____

[12] A detailed analysis of this incident was conducted by the United States Fire Administration as part of the Major Fires Investigation Project. The report, "Three Firefighters Die in Pittsburgh House Fire," is available for free from the United States Fire Administration.

Incident 13

On February 25 Chief Jimmy Bryant of the Indian Field (SC) Fire Department died of a heart attack after discovering a fire at a campground. The fire was deliberately set.

Incident 14

On March 3, Firefighter Neil Hyland of the Massapequa (NY) Volunteer Fire Department died in an automobile accident while responding to a fire call.

Incident 15

On March 5, Lt. Raymond Schiebel of the New York City (NY) Fire Department went into cardiac arrest while operating at a fire in Brooklyn. Schiebel died two days later at a hospital. An investigation into the incident revealed that a paramedic allegedly failed to properly intubate Lt. Schiebel during resuscitation efforts, inserting the endotrachial tube into his esophagus instead of into his trachea.

Incident 16

On March 8, Firefighter Donald Koebel of the Johnson County (KS) Fire District #2 died when he became trapped in the basement of a house fire and ran out of air. Koebel was part of the initial entry crew attempting to locate the seat of the fire when the floor collapsed beneath him. Heavy smoke and fire conditions prevented other firefighters from rescuing him. The fire originated in the basement of the house.

Incident 17

On March 9, Lieutenant Joseph Mambretti of the San Francisco (CA) Fire Department died of severe respiratory burns he received after becoming trapped in the garage of a house that was on fire. Mambretti, the officer on the first arriving engine, had led his crew with the first attack line into the garage when the electrically controlled garage door closed behind them. The fire spread quickly due to 50 mile-per-hour winds, creating heavy fire conditions in the garage and injuring the three firefighters before other crews could breach garage door with axes and saws to pull them out.

Incident 18

On March 13, Forest Ranger Bobby Crowe of the Georgia Forestry Commission died of a heart attack after battling a fire in a half-acre of wood pallets.

Incident 19

On March 15, Firefighter Phillip Sherburn of the Aumsville (OR) Rural Fire Protection District died of a heart attack shortly after responding to a house fire. Sherburn collapsed while performing water supply operations.

Incident 20

On March 18, Firefighter Henry Williams of the Delran (NJ) Volunteer Fire Department died when he suffered a heart attack while taking a firefighter stress test for the New Jersey Forest Service.

Incident 21

On March 24, Fire Engineer Donald Kaczka of the Chicago (IL) Fire Department died after suffering a heart attack at the scene of a rubbish fire at a recycling plant.

Incident 22

On March 27, Firefighter Dana Morrison of the Ferry County (WA) Fire Protection District No. 2 died after suffering a heart attack while operating a hose line at a fire.

Incident 23

On March 29, Deputy Chief Norman Prime of the South China (MA) Fire Department died of a heart attack while fighting a brush fire.

Incident 24

On April 2, Firefighter James Weaver of the Gallupville (NY) Volunteer Fire Department died of a heart attack while performing water supply operations at a brush fire. Weaver was 71 years old.

Incident 25

On April 13, Firefighter Herloff "Ted" Hansen of the Hobart (IN) Fire Department was killed while conducting search operations for two reported trapped victims at a house fire. Hanson and another firefighter were operating on the second floor when fire erupted from a concealed space near the stairwell, trapping them on the second floor, where they ran out of air. They were able to find their way to a window where a rescue ladder had been placed. Hansen aided his injured partner through the window and on to the ladder when a flashover occurred and he was killed. Three other firefighters were injured attempting to rescue Hansen and his partner.

Incident 26

On April 15, Forestry Technician James Wilson of the Tennessee Division of Forestry suffered a fatal heart attack after returning from the scene of a four acre brush fire.

Incident 27

On April 17, Engineer Judith Luster-Stauss and Firefighter Michael Lohbeck of the Castella (CA) Volunteer Fire Department were killed while responding to a barn fire when their tanker truck failed to negotiate a curve and overturned into a creek. Both died of traumatic injuries at the scene of the accident. The firefighters had apparently gone in the wrong direction and were reported to be heading away from the fire at the time of the crash.

Incident 28

On April 23, Lieutenant Earl McNeil, Jr. of the Princess Anne (MD) Volunteer Fire Department died of a heart attack after fighting a brush fire. Lt. McNeil was a retired career firefighter with the Boston Fire Department.

Incident 29

On April 24, Firefighter Leroy Cropper of the Ocean City (MD) Volunteer Fire Department suffered a heart attack after fighting a fire in a hotel. Cropper was hospitalized and died on April 28.

Incident 30

On May 5, Firefighter Greg Cusson of the Noble Township (IN) Volunteer Fire Department was killed when his car collided with another fire department vehicle at the scene of a reported explosion at a school.

Incident 31

On May 9, Firefighter Travis McCormick of the New Union (AL) Volunteer Fire Department suffered a fatal heart attack at the scene of a mobile home fire.

Incident 32

On May 12, Chief Ray Parnell McKay, Jr., of the Northeast Lamar County (MS) Fire Department suffered a fatal heart attack as he was leaving the fireground.

Incident 33

On May 12, Firefighter Dana Stivers of the North Pulaski (AR) Fire Protection District No. 15 was killed when the fire engine in which she was riding overturned, crushing her. The driver of the engine received minor injuries. The engine was on a non-emergency training run and the driver had apparently swerved to avoid an oncoming car. Stivers had just completed her six month probation period in the department.

Incident 34

On May 19, Robert Lapp of the Grantsville (MD) Volunteer Fire Department and the Northern Garrett (MD) Volunteer Rescue Squad died of an apparent heart attack while transporting a patient to the hospital from the scene of a motor vehicle accident.

Incident 35

On May 24, Firefighter Ron Deer of the Wayne Township (IN) Fire Department was killed when the fire engine in which he was riding overturned. One other firefighter was paralyzed in the incident; two others received minor injuries. Deer and the other critically injured firefighter were in the back of the engine and were not wearing seatbelts when the accident occurred. The raised roof of the engine separated from the cab when it overturned and they were thrown from the engine . The engine was responding on a box alarm which turned out to be a false alarm.

Incident 36

On June 3, Chief Bradley Hocking, Sr. of the Pipestone-Berrien Township-Eauclaire (MI) Fire Department suffered a heart attack after responding to the scene of a fatal vehicle accident. Chief Hocking was transported by an ambulance to a nearby hospital, where he died on June 6.

Incident 37

On June 5, Firefighter William Walls of the Rock Community (MO) Fire Protection District suffered a fatal heart attack after fighting a fire in a mobile home.

Incident 38

On June 6, Lieutenant Peter "Butch" Borwegan of the Edison (NJ) Division of Fire was discovered unconscious on the apparatus floor of Edison Fire Station 5. Attempts to resuscitate Lt. Borwegan were unsuccessful, and he was pronounced dead at a nearby hospital.

Incident 39

On June 7, Firefighter David Barrera of the Eagle Pass (TX) Fire Department died after suffering a seizure while on duty as a dispatcher.

Incident 40

On June 10, Firefighters Richard Hogan and Christopher Rizac of the Sheppard Air Force Base (TX) Fire Department were killed while fighting a fire at an oil refinery in Addington, Oklahoma. Hogan and Rizac had responded in a P-19 crash truck to assist local fire departments with suppression efforts. The fire was located in a oil storage tank and had been caused by a lighting strike. Rizac and Hogan were killed when several thousand gallons of burning oil boiled over the side of the tank, trapping their crash truck in a mixture of oil and mud. They attempted to flee on foot but were overrun by the flow of oil and died of massive burns.

Incident 41

On June 12, Fire Commissioner Kevin Sutch of the Levittown (NY) Fire Department suffered a fatal heart attack while attending the New York State Firemen's Convention in Albany, New York.

Incident 42

On June 22, Contract Pilots Gary Cockrell and Lisa Netsch of Aero Union Corporation, and Pilot Michael Smith of the USDA Forest Service were killed in an aircraft collision over Ramona, California when Smith's spotter plane hit the DC-4 air tanker piloted by Cockrell and Netsch. Both planes were on final approach to the airport after returning from dropping fire retardant on a brush fire.

Incident 43

On June 22, Assistant Chief Carter Martin of the Brookville-Timberlake (VA) Volunteer Fire Department drowned after he waded into fast moving flood waters to search three vehicles that had been swept downstream. Chief Martin was swept under water and trapped. He was wearing full protective firefighting turnout clothing while in the water, was not wearing a life vest and had been tied into a rope that was attached to a fire engine. Two other rescuers were thrown into the water but survived. Martin also served as a career firefighter with the Lynchburg (VA) Fire Department and was an instructor with the Virginia Department of Fire Programs. It was later discovered that the occupants of the vehicles had reached safety on their own prior to the response of emergency personnel.

Incident 44

On June 29, Fire Protection Specialist John Woodward of the New York State Office of Fire Prevention and Control died of a heart attack while conducting a fire inspection.

Incident 45

On July 6, Lieutenant Randy Williford of the North Little Rock (AR) Fire Department suffered a heart attack after attempting to complete an agility test required for a promotion to the rank of Captain. Williford died of heart failure at a hospital on July 9.

Incident 46

On July 10, Firefighter Gary Soupene of the Riley County (KS) Rural Fire Department was killed while responding in his personal vehicle to a reported grass fire. Soupene was slowing down to pick up another volunteer when his vehicle was struck from behind by a car driven by another firefighter responding to the same incident, flipping Soupene's car and killing him. The reported fire turned out to be a bale of hay that had caught on fire.

Incident 47

On July 10, Firefighter John Schuyler of the Weldon (PA) Fire Company suffered a fatal heart attack while responding on foot to the fire house after a vehicle fire had been dispatched.

Incident 48

On July 14, Firefighter John Weingart of the Detroit (MI) Fire Department suffered a fatal heart attack while hooking up to a fire hydrant at a residential dwelling fire.

Incident 49

On July 15, Assistant Chief Edward Pitcher of the Sharon County (CT) Volunteer Fire Department was electrocuted when he came in contact with a downed power line during the clean up of debris after a storm.

Incident 50

On July 15, Adam Sorenson of the Ruth (NV) Volunteer Fire Department was killed when the ambulance he was driving swerved and crashed. The ambulance was not responding to an emergency or transporting any patients when the accident occurred.

Incident 51

On July 19, Mayor Arthur Thompson, Fire Commissioner for the Freeport (NY) Fire Department, suffered a heart attack while en route to a fire. He was transported by an ambulance to a local hospital where he died.

Incident 52

On July 20, Firefighter Lyle Garlinghouse of the Osceola (FL) Fire Department suffered a fatal heart attack at an EMS incident.

Incident 53

On July 21, Firefighter Peter Crown, a helicopter pilot for the Honolulu (HI) Fire Department was killed when his helicopter crashed in the Koolau Mountains on the island of Oahu. Crown had been conducting a search for a lost hiker and had been towing two police officers in the helicopter's basket to the search area when the helicopter crashed in inclement weather. The two police officers were also killed.

Incident 54

On July 25, Firefighter Mitch Weaver of the Tunnelton (WV) Volunteer Fire Department died of a heart attack while at the scene of a vehicle accident.

Incident 55

On July 28, Firefighters Bill Buttram and Josh Oliver of the Kuna (ID) Rural Fire Department were killed when the 1955 brush truck they were driving stalled and they were overrun by a fast moving wildfire.

Incident 56

On July 30, Chief William Luker of the Cedar Creek (MS) Volunteer Fire Department was killed when the 1972 3,500 gallon tanker truck he was driving overturned en route to a barn fire, ejecting him from the truck. Luker was not wearing a seat belt at the time of the accident.

Incident 57

On August 1, June Fitzpatrick of the Rocky Point (NY) Fire District died of a stroke while en route to the fire station for an emergency.

Incident 58

On August 5, Firefighter William Marks of the Munhall (PA) Fire and EMS 1 was killed when the engine on which he was riding jumped a curb and overturned en route to an electrical fire in a row house. Marks was riding on the tailboard of the engine at the time of the accident. Several other firefighters were seriously injured.

Incident 59

On August 7, Firefighter Eric Mangieri of the New Kensington (PA) Fire Company was killed when he became trapped while trying to escape from a house fire during a flashover.

Incident 60

On August 16, Firefighter Christopher Garneau of the Warrenton (VA) Volunteer Fire Company died of a heart attack while responding to an emergency incident. An autopsy revealed that Garneau, age 17, had an enlarged heart.

Incident 61

On August 21, Firefighter-EMT Bruce Cormican of the Black River Falls (WI) Fire Department drowned while conducting a body recovery in a creek for a drowning victim. Cormican and two other members of the Black River Fall dive team were trapped in a hydraulic created by a small waterfall while wading into the creek and searching for the victim's body. Shore personnel were able to remove the three rescuers with a pike pole, but they could not revive Cormican after pulling him out of the water.

Incident 62

On August 24, Rescue Diver Corey Berggren of the Knoxville (TN) Volunteer Rescue Squad died while conducting dive operations in a quarry to recover the body of a drowning victim. Berggren had apparently breathed the wrong mixture of gas from his SCUBA tank during a the 200 foot dive, asphyxiating him.

Incident 63

On August 27, Firefighter James Greg Hinson of the Mebane (NC) Fire Department drowned after rescuing a man from a vehicle on a flooded highway. Hinson and two other firefighters had tied themselves into a rope that was attached to a haul line on a fire department vehicle. As he reached the car and pulled the occupant out, the car was swept into the flooded creek channel. Hinson and the other firefighters slipped into the current, and he became trapped on a guide wire to a telephone pole when other rescuers tried to haul them out of the water. The two other firefighters were rescued. A rescue diver from another department entered the water to reach the car driver, and they were both swept into a tree, where they were later rescued by boat. Hinson was pulled from the water within seven minutes in cardiac arrest.

Incident 64

On August 31, Firefighter Marty Kautz of the Brush (CO) Volunteer Fire Department was killed in a vehicle accident while en route to a medical emergency. Kautz and three other firefighters responded in his minivan from the fire station to a report of a child choking. They were struck en route by a car driven by an emergency medical technician who had run a stop sign while en route to the ambulance quarters to respond to the same call. Kautz was killed and the three firefighters injured.

Incident 65

On September 6, Fire-Police George Peters of the Eflinwild (PA) Volunteer Fire Company suffered a fatal heart attack at the scene of a multi-alarm fire.

Incident 66

On September 8th, Past Chief John Pache of the Aviation (NY) Volunteer Fire Company No. 3 suffered a fatal heart attack while en route to a structure fire.

Incident 67

On September 13, Captain John McCroden of the City of Geneva (OH) Fire Department died when he suffered a heart attack while ventilating the roof at a fire in a single family house.

Incident 68

On September 16, Firefighter Eric Schaefer of the Baltimore City (MD) Fire Department was killed when a 1 1/2 foot thick granite wall collapsed on him while he was engaged in forcible entry at a multiple alarm fire in an old warehouse that had been converted to numerous shops and businesses.

Incident 69

On September 17, Firefighter Gene Schubert of the Harriman (TN) Fire Department suffered a fatal heart attack at the scene of a fire.

Incident 70

On September 19, Captain Ray Leccioni of the Colma (CA) Fire District suffered a fatal heart attack while at a medical call.

Incident 71

On September 23, First Assistant Chief Frederick Fairweather of the Bullville (NY) Fire Company suffered a fatal heart attack while marching with the department in the Bullville parade. He was a retired firefighter with the Newburgh Fire Department.

Incident 72

On September 26, Lt. Thomas O'Boyle of the Chicago (IL) Fire Department suffered a fatal heart attack at a fire in a furniture warehouse. Lt. O'Boyle, aide to the Fire Commissioner, had just exited the structure after checking on the progress of firefighters inside when he collapsed.

Incident 73

On September 28, Sergeant John Fisher of the Greensburg (PA) Volunteer Fire Department suffered a fatal heart attack while responding on foot to an emergency call for a pedestrian struck by a train.

Incident 74

On September 30, Instructor Richard Washburn of the Kentucky Tech Fire/Rescue Training, and a member of the Whitehall (KY) Fire Department, collapsed and died while he was teaching a confined space entry and rescue course at a regional fire school.

Incident 75

On October 8, Firefighter Peter McLaughlin of the New York City (NY) Fire Department was killed while performing a search on the fourth floor at a fire in a tenement building, when the fire broke through the ceiling engulfing the apartment in flames. McLaughlin's route of escape was blocked by a window gate. He died of burn injuries and smoke inhalation. The fire started on a mattress in a fourth floor bedroom. The building had been previously cited for over 170 fire code violations.

Incident 76

On October 26, Corporal John Riggins, Jr. of the Indianapolis (IN) Fire Department suffered a fatal heart attack after performing roof ventilation at a house fire.

Incident 77

On October 28, Past Chief James "Frank" Ainsworth of the Friendship (WV) Fire Company suffered a fatal heart attack at a fundraising event for the fire company. Ainsworth was treasurer and served on the Board of Directors of the department at the time of his death.

Incident 78

On October 29, Firefighter Stephen Sulzinski of the Hicksville (NY) Fire Department suffered a fatal heart attack while attending a fire service activity in Albany.

Incident 79

On November 7, Asst. Chief Walter Augustin of the Congers (NY) Fire Department suffered a fatal heart attack while marching in the Congers parade with the department.

Incident 80

On November 9, Firefighter John Haviar, a industrial fire brigade member at a Reynolds Aluminum plant, was killed when he entered an oxygen deficient atmosphere in an excavation pit without breathing apparatus to attempt the rescue of three workers. Argon gas had been accidentally pumped into the pit, displacing the oxygen and trapping the workers. Haviar was overcome and was killed along with two of the workers.

Incident 81

On November 11, Chief Thomas Buff, Jr. of the Blaney (SC) Volunteer Fire Department was killed when he was struck by a passing police cruiser. The Blaney VFD had just finished extinguishing a vehicle fire on an interstate highway when a minor accident occurred in the opposing lanes of traffic. Chief Buff started across the road to check on the occupants of the cars when he was hit, receiving multiple traumatic injuries.

Incident 82

On November 19, Assistant Chief David Harness of the Hanna Township (IN) Volunteer Fire Department was killed when he was struck by a vehicle at the scene of an emergency.

Incident 83

On November 25, Firefighter Michael Canonico of the Andover Township (NJ) Fire Department was killed while responding on his motorcycle to a report of a furnace fire. Canonico died when he attempted to pass a pickup truck on the right and the truck made a right turn, striking him.

Incident 84

On December 14, Captain James Shue of the Locke Township (NC) Volunteer Fire Department was killed when the engine he was driving overturned en route to an odor of smoke call, which turned out to be a false alarm. Two other firefighters were injured.

Incident 85

On December 31, Lt. John Clancy of the New York City (NY) Fire Department was killed when the floor collapsed beneath him at a fire in an abandoned residential building as he entered to conduct a search for occupants. Lt. Clancy fell into the basement where he died of burns.

`